The INVENTION *of* MIRACLES

Katie Booth

The INVENTION *of* MIRACLES

language, power, and Alexander Graham Bell's quest to end deafness

SCRIBE
Melbourne • London

Scribe Publications
2 John St, Clerkenwell, London, WC1N 2ES, United Kingdom
18–20 Edward St, Brunswick, Victoria 3056, Australia

Published by Scribe 2021

Interior design by Ruth Lee-Mui

Printed and bound in the UK by CPI Group (UK) Ltd, Croydon CR0 4YY

Scribe Publications is committed to the sustainable use of natural resources and the use of paper products made responsibly from those resources.

9781913348403 (UK hardback)
9781922310323 (Australian paperback)
9781925938746 (ebook)

Catalogue records for this book are available from the National Library of Australia and the British Library.

scribepublications.co.uk
scribepublications.com.au

For

Harry McCarthy, Rose McCarthy,

and Rita Cyr

Contents

PART III

On Miracles

In 1679, the first recorded attempt to teach speech to a deaf child in America was cut short by the local church, due to concerns that the teacher was committing blasphemy by trying to perform a miracle.

Two hundred years later, Alexander Graham Bell's own work to teach the deaf to speak would be called miraculous, over and over again.

But a miracle hinges on what remains hidden.

Prologue

In the hospital bed, my grandmother faced the window, bathed in the bluish light of the rising moon. I watched her body, memorizing it: the wrinkles of her fingers, the way her jawbone tucked back into her neck, the way her tongue moved like a soft oyster in the shell of her mouth. I had known her my whole life, but I had never before seen her weak. Now it was hard to look at her. The roundness of her body, which once seemed so soft and warm, was only heavy, seemed only to pin her to that hospital bed. Her usual facial expression—eyes sharp, mouth firm—was now tired, resigned. Her head lay on the pillow; her gray hair, thin and oily, was swept back from her face; her eyes were blank and dark. She wasn't dead, not yet, but something had shifted.

Normally I saw my grandmother in her home, among her deaf community, where apartments were set up with blinking lights for alarms, where telephones released little scrolls of typed English, and where furniture was arranged for open sightlines in order to use American Sign Language (ASL) across rooms. She lived among the culturally deaf, defined by the use of ASL and observation of deaf cultural norms. In those spaces, my grandmother had more access to information than I did. With friends she communicated in quick, fluent ASL, and even when I could catch the gist of the individual words I could also tell that there were layers and layers of meaning that were escaping me. They were carried in a small twitch of an eyebrow, the subtle lift of the corner of a lip, an invasion of space, or a quick shift away. In her world, she was firm, strong, steady. But here in the hospital, things were different.

My grandmother had suffered her heart attack four days before and had been in the hospital, alone, for the three days that followed. Only then did anyone get in touch with our family. In the meantime, my grandmother's presence barely seemed to have registered. My grandmother, through notes scrawled in English, had made several requests—few of which seemed to have been addressed. She asked for someone to contact us, and for a TTY, a text-based telephone device, so she could call us herself. They gave her a TTY but ignored her insistence that it was broken. My grandmother had asked for an interpreter at least four times before we arrived, and received one only once, when the cardiologist came to see her. Even then, she misunderstood her diagnosis—she had no idea of the severity of the situation, the damage that had been done to her heart. With almost no information, she went on waiting for us. For those three days, we had no idea that she was lying there.

I was nineteen at the time and knew that a deaf family member in the hospital was an all-hands-on-deck situation. When I got the call, I took a late-night Greyhound back home from college and accompanied my mother to the hospital the next morning.

In this environment, I had access that my grandmother didn't, and the fact sat uneasily in my stomach. As I watched my grandmother, I listened for my mother's voice as she spoke to the doctors and nurses in the hallway. I couldn't hear her sentences, but I could hear the way her sounds began gently and then rose firm in advocacy. My mother knew the hospital staff would listen to her because they always turned to the hearing family members in these moments. They wrongfully saw us as the interpreters, caretakers, decision makers. Without us, they behaved as though there was simply no one with whom to communicate.

It happened everywhere. When strangers realized my grandmother was deaf, their faces contorted with discomfort or gawking fascination. Their bodies became stiff, moved away from her. They made mistakes. At restaurants, waiters were too flustered to serve her. When they did, chances were small that she would get what she had asked for, even though she knew how to point clearly at items on a menu and write any additional requests in her memo pad for them to read. At the mall, cashiers always forgot to remove security tags, and so she always set off alarms that she couldn't hear. She was chased down by security guards, who placed their hands angrily on her shocked arms before she had any idea what was happening.

Here, at the hospital, doctors and nurses did what they pleased with her body. Often they didn't look at her face at all. They avoided her eyes, which were hungry for information and seemed to embarrass them. If they spoke to her, they held their eyes big and moved their mouths in long, round shapes. They seemed to think that their distorted mouths could substitute for a certified interpreter, but my grandmother could make little meaning out of the charade. They ended with saccharine smiles, like everything was okay now, and then tugged at her arms, stuck needles into her veins, or rolled her bed to the operating room for open-heart surgery. Mostly, they spoke to one another. Their faces hovered intermittently above her and then, always, they turned away.

When my grandmother tried to communicate, she was treated as a bother, but when my mother or I spoke, hospital staff listened. Now my mother was harnessing this fact to resist its inherent inequality. From my grandmother's room, my ears strained to hear her. I could capture phrases like *Americans with Disabilities Act* and *informed consent* and *civil rights*, legal code words that culminated in *violation of federal law*. She was battling for my grandmother's right to access her own language, her right to understand and to be understood.

By this point, the early 2000s, the broader deaf community had been entrenched in this fight for nearly 150 years. They'd been fighting ever since one of the foremost teachers of the deaf began to say that ASL was unnecessary, that the deaf could do without it—and should be forced to, if possible. In the hearing world, this teacher is renowned, but not for the generations-long struggle he began. He's known as a plucky inventor, a man who helped people to communicate across cities, across nations, across oceans. In the deaf world, however, he's remembered with rage. He's the man who launched a war in which the deaf would have to fight for their lives.

I could hear my mother's voice as the doctor pulled her aside for privacy, and then the doctor's low murmur. When my mother responded, I could hear no language at all, just soft undulations of urgency. I measured the wreckage by the tones of her voice. I looked at my grandmother's face, lit evening blue, and counted her shallow breaths.

When I was a child, I always thought of my grandmother as strong, sturdy, immovable. She carried her dignity like she carried our language: on her body. She stood tall and with her chin high, scanning the world around her. She clutched her purse in one hand, my hand in the other. We went out like that—steeled, lips tight—into the hearing world. But when we needed to communicate with each other, we could shrug all that rigidity off; we would slip our hands free to talk.

Our language took our full attention, our full bodies. We leaned forward into certain words, backward into descriptions of things that had happened before. When my grandmother and I signed, we arranged our words in the space before us and protected the space of our words. Our lips brought almost no sound but still moved with meaning, letting loose little clicks and puffs. In ASL, my grandmother was firm, direct, and controlling. When she summoned warmth it looked a little like she was faking it, like she was trying out the feel of a smile.

She didn't show warmth on her face so much as she showed it through small gestures. After my grandfather died, she moved out of the little house down the street and into a deaf apartment complex about an hour away. Our visits became less frequent then, and this care she exhibited took on a new intensity. She was always making me cinnamon raisin toast and gave me small gifts whenever I visited. If she had forgotten to get one, she would pluck a mirror from her wall.

"I bought this for you," she'd say, as though it didn't leave a white shadow where it once hung. "I just put it there in the meantime."

If I said I liked something, she wouldn't let me leave her home without it. "It's for you," she'd say, pushing it sternly into my arms. There was no arguing with her; she was forceful even in her generosity. If I wasn't careful, I could have a whole loaf of bread put before me, slice by buttered slice. By the time I was a teenager, I had learned to compliment her belongings selectively, lest I leave her shelves and walls bare.

When I was young, we mostly communicated through ASL; as I grew, and began to forget the language, this was supplemented with written English. Occasionally, my grandmother summoned her voice, but since she didn't use her voice with strangers, it always felt special, something that only came out when things were safe. Sometimes her voice made words in English, and sometimes her voice was just a thing that arose from her, something that captured a particular mood or mimicked the vibrations she felt in the world. It communicated on its own terms. Staticky and monotone and often indecipherable to those outside our family, my grandmother's voice is one of the sweetest sounds I've ever known.

As a child it never occurred to me that these elements couldn't coexist—speaking and signing—nor did I know that the demand placed on voice was one of the most painful things that happened, and continues to happen, to deaf people. In the hearing world, it would come to represent their very humanity; deaf people who couldn't speak were often referred to as monkeys, or prehumans. I didn't know that the chief person behind the campaign to keep deaf children from learning ASL was the man whom most people thought of differently, pleasantly, as the inventor of the telephone. Or that the movement he led would forever change what was expected of the deaf. Alexander Graham Bell, a man who had made so-called miracles out of voice, was at the center of it all.

I sat beside my grandmother in the hospital, where her humanity was being ignored. My mother's voice—*violation of federal law*—lingered in the air, but it felt like legal protections mattered little. We still had to explain yet again that the deaf were legally and ethically entitled to the free use of their language. That my grandmother had the right to know why she had been hospitalized. That she had the right to contact us. That she was human, and her language was real and true and necessary and wholly distinct from English. It was never more vivid to me that until we could usher in a widespread understanding of ASL as a dignified and necessary language, deaf culture—and deaf lives—remained at risk. This theft of language, culture, and dignity was the legacy of Alexander Graham Bell.

I grew up between English and ASL, between the worlds of the hearing and the deaf. My maternal grandparents lived in a home we called Knotty Pine, just around the bend from my house. At home, my parents, both hearing, spoke mostly English with a smattering of ASL, but when they went to work, I went to Knotty Pine, where I was immersed in deafness.

In my grandparents' home, we didn't speak, since we didn't need to; we needed neither speech nor its accompanying language, English. Instead, our eyes were trained to respond to the slightest movements in our periphery. We scanned the visual world for both information and language, and when we spoke we did so with our hands and our bodies. Our language was intimate—it was private, something most people didn't know; it required close attention to each other's faces, and so we spent our conversations looking into each other's eyes; and since it doesn't lend itself to multitasking, when we were doing something else, we filled those moments not with the clatter of speech but with the tenderness of observation, of taking each other in.

Of course, I was only a child, and much of this was lost on me. It was simply the way we lived. I didn't think much about my grandparents' deafness, but there were mysteries inside of it that grabbed me even then. My grandmother had told me countless times that she couldn't hear anything, and I mostly believed her, but I was also fundamentally confused. I thought her deafness applied only to outside sounds, like my voice or the honk of a car. For some reason I still thought she could hear her own body: the gurgle of her stomach or the hum of her voice before it was released. I thought that she could hear herself, shimmering inside, like I could. I always imagined that this is how she learned to speak.

I wasn't completely wrong. There is a shimmering, vibration, and sensation that spreads out from the speech organs that she could use to understand the sounds she made. But she couldn't hear herself the way I understood hearing. We were sitting on her screened-in porch when she told me again that she couldn't hear anything at all. Not even herself. Someone taught her to speak anyway.

My brows furrowed in. In ASL, this indicates an open-ended question, even without the hand movements for *why*, for *how*. My grandmother bent her body down to the floor to reach for one of my coloring books and tore off the corner of a page. She sat back, held the torn slip in front of her lips, said *puh*, and the slip of paper blew back.

"That's how," she said, as if that explained everything.

It explained nothing. Instead, it opened a breathless rush of questions inside of me, but before I could ask anything, my grandmother saw the uncertainty begin to gather on my face. She sighed and stood.

"Wait here," she said, and she went into the house, where I could hear her rummaging around in the kitchen. When she returned, she held a birthday candle between her pointer and thumb. She lit it and held it hot in front of my face.

"Say *puh*," she said.

I did, and the flame flickered in my breath.

She brought it up to her mouth, said *puh*, and blew it out. "Like that," she said.

I still didn't believe her; it didn't seem like this could be a way to learn to speak. Besides, she couldn't always be trusted: she was the type of woman who cheated at Go Fish and then refused to look up so that you couldn't argue with her. Now her eyes pressed into me, and I nodded, complicit in her con. I could understand how the flame could help her say the letter *P*, but what about all the other letters? What about the *mmmms* and the *shhh*s? The *grrr*s and the *aaahhh*s? I thought it was just another one of my grandmother's tricks.

Later that day, when I asked my mother about it, she confirmed what my grandmother had told me. She explained when my grandmother went to school in the thirties, at the Beverly School for the Deaf in Massachusetts, teachers would do whatever they could to help the students *see* the sounds so that they could understand whether they were making the sound the right way or the wrong way. A student who pronounced *muh* instead of *puh* couldn't blow out a candle. And from that simple notion of learning to see sound unraveled a cultural story that would follow me for years.

My mother told me that long ago people thought that sign language was bad, and they punished the children who used it in school. The teachers slapped the students' palms or made them sit on their hands. This is what happened to my grandmother's whole generation—my grandmother, grandfather, great-aunts and -uncles—they were all deaf; they were all punished. This education was called oralism, my mother said, and it hurt our family very much. Everyone's hands were bruised.

Even as a child I understood this complex violation. My grandmother's hands were precious. They were a central element of her language; an attack on the hands was an attack on identity. I couldn't imagine my grandmother's hands as they might have been when she was young, and I couldn't imagine them bruised, either. It hurt just to think about it, and I shook off the image. I wanted, instead, to understand why. Why would anyone do this?

My mother explained that what the oralist teachers wanted instead of our language of the body was for deaf children to speak. They thought that if the children could speak and lip-read, then they could appear to be hearing. They thought that hearing was better.

This, too, was a new concept. It seemed to me that the deaf had a set of skills that verged on superpowers: they noticed things I never saw; they could tell immediately if I was lying; so attuned were they to the world that I regularly questioned whether my deaf relatives actually *could* hear.

But these teachers didn't understand deafness the same way as we did, my mother explained. They thought hearing was better and that the skills of speech and lip-reading would help the deaf. The deaf still wouldn't have the actual ability to hear, but they would be socially equal.

This was Alexander Graham Bell's grand plan, a hopeful dream that went very wrong, as his devotion to this miraculous idea eclipsed real social advancement for deaf people. This was the story of how ASL became hated, of how people came to see the signing deaf as stupid or weak or not quite human at all. Ultimately, it was the story of seeking to end deafness altogether—the language of the deaf, and the people, too.

But how anyone could see ASL or deafness as a problem made no sense to me. ASL made my family members stronger, smarter, more themselves. Many of the members of my deaf family could interact in the hearing world using English, writing messages on the tiny memo books they always carried, but I knew that their bodies were different when they did that, rigid. English wasn't *bad*, but it wasn't quite *theirs*, either. I understood that my family members were still the same people in English as they were when they signed, but also that they weren't. They were shadows of themselves; without the free use of their language, they were hidden inside their own bodies.

When we were with only each other, everything shifted. They were noisier without so many hearing people around, free to bang on tables and stomp the floor, sending vibrations designed to capture attention. My grandparents' voices released, with all their rough edges; unbridled by the convention of words, my grandparents clucked to their cats, moaned their yawns, yelped surprised laughter. When voice wasn't wrapped up in shame, it came as naturally to my grandparents as it does to anyone. And in the safety of the deaf world, their bodies became their language again.

They communicated between the kitchen and living room effortlessly—there were no walls there to impede their words, and my grandmother moved easily around their small kitchen, signing to my grandfather from behind the counter and piling two small plates with torn pieces of cold roasted chicken. My grandfather, one eye on the TV, signed to her with his careful gentleness; she responded with her sharp signs, always with an unpredictable edge. But then she'd bring him his plate of chicken and sit down with her own, each of their armchairs facing the TV and me on the floor between them. Everything soft and easy.

When they told stories, their faces took on the features of the people they talked about. My grandfather would make a stubborn purse of the lips to mimic my grandmother; he'd have a gentle clumsiness with language for my father, who had only begun to learn to sign after meeting my mother. We had special names in that language, too. Mine was the letter *k*, swept twice from the corner of the eye to the temple. My mother was a little waving *b*. My grandfather a double tap of an *h* at the shoulder, a loose allusion to the sign for *strong*.

Our bodies took up all the space before us. We positioned imaginary people and objects in the signing area to tell a story, shifting between them with a pivot of the shoulders or an expression of the face. My grandmother, her lips pursed, stood *here.* My father, the clumsy signer, stood *there.* Clumsy Signer said *this*. Pursed Lips's face transformed: her eyes big, her mouth agape. She couldn't believe what he'd said—he was *so horny*? But then she remembered a common hearing mistake. *Hungry.* He was so *hungry*. Clumsy Signer turned red, bashful and confused. As she showed him the correct sign, *hungry*, her face was stern—it was especially important for him to get this sign right.

Our words were everyplace at once. A spatial orchestra, they lived in all the dimensions of our language—our faces added layers of meaning; our hands moved forward or backward to help establish tense; our fingers described the characteristics of a stare, the speed of walking, a collision of forces, a gentle show of love. It seemed as clear as anything that resisting ASL meant resisting something pure and good and liberating, meant resisting the full personhood of the people I loved.

But not everyone believed this. Toward the end of his life, Bell had said that the signing deaf "represent our failures. Let us have as few of them as we possibly can." These people were my great-grandparents, my grandparents, my great-aunts and -uncles. Because of Bell's influence, they learned to be ashamed because they used ASL. For their whole lives, their joy in free communication, so easily granted to hearing people, was tainted by this shame.

Children in oralist schools were supposed to use English, and only English. Speak English, lip-read English, read and write English. It took priority over all other learning; often it entirely displaced other learning. Deaf children who signed, even accidentally, were scolded, mocked, denied social outings, and sent to the head matron, the disciplinarian. Their hands were hit. They were given milk and bread for dinner.

In the towns around these schools, the locals knew that if they saw children signing, they should report it—and they did. My great-aunt Rita went to the Clarke School for the Deaf in the late thirties and into the forties, where she was a residential student. When she traveled home for vacations, she would sign with her all-deaf, all-signing family. When she returned to school, her teachers said they could tell she had been signing. She was punished.

The scars of these educations are deep; they are real. My great-aunt Rita rarely used her voice when I knew her; she maintained that her speaking peers were on a "higher level," and her own voice, she said, was ugly. When she laughed, she did so silently—this is what the schools taught deaf children whose laughter was too unbridled, too loud, or too free. When her husband, my great-uncle, died, we tried to call her, but she wouldn't respond. A teacher had once told her that she sounded like an animal when she cried, and so she cried alone.

On his deathbed, my grandfather told my mother that he wouldn't be going to heaven. "God doesn't love me," he said, "because I am stupid; because I am low. Because I sign."

This happened when I was seven, but I wouldn't learn about it until years later. By then I knew whom to blame; by then I had learned that Bell's legacy went even deeper, even darker than I'd known. When we covered the Holocaust in school, I learned that Hitler had targeted the Jews. At home, I learned that Hitler also targeted the deaf and the disabled, and that decades before Hitler, at the very beginnings of the eugenics movement, there was Bell. He wanted the deaf eradicated, their marriages to each other forbidden, their procreation ceased. I understood this: if Bell had had his way, my great-aunt and great-uncles wouldn't be here, my grandmother wouldn't be here—and by extension, my mother wouldn't be here, and I wouldn't be here.

So when my grade-school social studies book said that Alexander Graham Bell was the inventor of the telephone, it sounded as absurd to me as introducing Adolf Hitler as a vegetarian who once ruled over Germany. Bell didn't just want my family's language and culture to be obliterated—he wanted us as a people to be obliterated. I didn't understand how he could arrive at that idea, and even less could I understand how he could get away with it.

Nevertheless, there he was: a black-and-white photo of a man with a white beard and bright black eyes; a man who looked dignified, controlled; a man who represented all hearing ignorance and violence against the deaf. It was clear to me as a child, and then later, looking down at my grandmother's body: Bell was responsible not only for his own actions but also for the actions of the inheritors of his mind-set.

At my grandmother's home in a deaf apartment complex, her best friend, Charlie, had slipped a note under her door. "People are asking where you are," he wrote. "I tell them Rose met a handsome billionaire and has retired to Florida."

Over the next few days, the hospital would continue to fight against getting an interpreter; they asked me to interpret complex medical documents for my grandmother, even though I couldn't even understand them in English; and it came to light that they had insisted my grandmother sign medical forms that she couldn't understand. Eventually, we were told that my grandmother needed a surgery that they couldn't provide, and that she needed to be transferred. The new hospital had a better department of cardiology, but my mother couldn't shake the look on the face of my grandmother's doctor at Beverly. When my mother had used the phrase *informed consent*, it had scared him. His eyes, so big.

By the time my grandmother got to the new hospital, they said she was far too weak for the surgery. We wondered if she was transferred to get rid of us. We wondered, too, if the transfer had made her weaker.

On the last day I saw my grandmother alive, she slipped her rings from her fingers and folded them into my and my sister's palms with her papery hands. Then she removed her wedding ring and tried to take my mother's hand, but my mother shook her head, closed her palm tight. My mother's and grandmother's eyes locked on to each other, becoming a closed circuit of stubbornness and heartbreak. Then my grandmother shook her head and shrugged, pushing the ring to my mother.

"They'll just steal it," she signed, side-eyeing the hallway as her fingers formed the word for *steal*: the index and pointer fingers, crooked, snatching the air.

"Take it for now," she signed, and offered the ring once more. My mother nodded, and our hands stilled.

I slipped the ring my grandmother had given me on and off each of my fingers as her breathing slipped in and out. I would spend years replaying these days in my mind, but in that moment I was only watching her breath, praying for her to breathe again. My grandmother was so fierce, and this seemed like a particularly cruel end to her life, this series of medical neglects, this asking and not receiving, this waiting with no answers, with hardly any words at all.

The next day, she died.

The events surrounding my grandmother's death made no sense to me, and so I couldn't file them away, couldn't move on from them. My grandmother was an inherently powerful woman, yet her power had been stripped from her. The root of her power—her language and culture—seemed to have undermined her ability to advocate for herself. There was some part of myself that wondered, in what I immediately knew to be the deepest betrayal, if all this could have been better if only my grandmother's body were different, if only she weren't deaf. At that moment I could hear his quiet voice inside of me, the voice of Alexander Graham Bell.

I understood that it was not her language that hurt her, but the reception of her language by the people who held power as her life neared its end. But at the time that seemed like a mere technicality. My grandmother's life had depended on those people.

The institution meant to care for her had put her into this broken situation. But I also understood that what had happened in the hospital couldn't be extricated from a broader web of institutional oppression, including the law, health care, employment, and, most fundamentally, education. The values of oralism had trickled into all hearing-dominant structures, sustaining all manner of violence against the deaf. Suddenly, I felt like I needed to know more about everything my mother had told me over the years, about the campaign against sign language, about how it shaped my family and how it shaped me. To do this, I had to learn about the man who had ushered this movement into the realm of the possible. Bell may have been dead for almost a century by then, but his ideas were as alive as ever.

About a year after we buried my grandmother, in the dead of winter 2004, I visited the Clarke School for the Deaf in Northampton, Massachusetts. It was the oralist school that many of my deaf family members had attended—not my grandmother but my three great-uncles and my great-aunt Rita. In the Clarke School archives, slowly, I began to learn about Bell—not the metaphor but the man.

I hadn't known, for example, about Mabel Hubbard, the woman who would become Bell's wife. I hadn't known that she was deaf and that her childhood was entwined with the founding of the Clarke School. I didn't know that her family didn't want her to marry Bell but that he fought for her right to decide for herself. I hadn't known that Bell's mother was also deaf, and that she had resisted her son's marriage to a deaf woman, fearing that they would have deaf children. But Bell shrugged her off, too. I hadn't known that the telephone had sprung from his work with the deaf, or that it was, in his mind, a distraction from his larger mission: oralism. And when his telephone investors threatened to pull out because he was spending too much time teaching, he refused to back down. Deaf education, he insisted, was his life's work. In the beginning, he didn't sound like the eugenicist he would become.

I wanted to know what Bell's contemporaries in the deaf world thought, and so I found an old library copy of Albert Ballin's diatribe against oralism, *The Deaf Mute Howls.* Its opening lines lodged deep inside of me: "Long, loud and cantankerous is the howl raised by the deaf-mute! It has to be if he wishes to be heard and listened to. He ought to keep it up incessantly until the wrongs inflicted on him will have been righted and done away with forever." This made sense to me, but I was surprised later in the book, when Ballin described his friendship with Bell. He wrote that "there never existed a more gracious, more affable, more fun-loving or jollier fellow on this sorry old globe than Dr. Bell. His character was without blemish. . . . His love and sympathy for the deaf were boundless, and should never be questioned."

But how could a deaf man with such revolutionary fire and pro-sign language allegiances be friends with Bell? There was something I didn't understand, both about Bell and his tangle of ideas and influences. I began to realize that Bell—like the hospital staff, like the teachers of my grandmother's childhood, like so many hearing people—probably set out wanting to help and to heal and to empower. This didn't erase the harm he did, but it did raise a bigger question, a question with implications not only for Bell's legacy but for so many of us: How does a man set out with such ferocity of love and work and end up using his place of power to advocate for ethnocide? He had good intentions, sure, but could "good intentions" really absolve him of the harm he wrought?

In order to answer these questions, and to understand how I—and hearing people everywhere—could avoid making the mistakes Bell made, I knew I would need to center my attention not on the telephone but on Bell's own self-professed life's mission—to teach the deaf to speak. Over the next fifteen years, I dug into everything I could find: Bell's private letters and his wife's diaries, nineteenth-century deaf newspapers and contemporary deaf studies journals, and archives ranging from the Clarke School Library to the Library of Congress to find out who Bell really was.

I did not expect to find a coming-of-age story, a romance, and a fall from grace along the way. I did not expect that I would grow strangely fond of Bell, nor that my anger at him could rage even deeper. I guess I never expected to understand him—not in any real way—or that in my quest to understand him, I would also learn painful truths about myself. His legacy, after all, has fundamentally shaped the way deafness is viewed in the hearing world, and I live in that world. I had set out wanting to burn Bell's legacy to the ground, but to do that would be to turn away from him, to pretend there were no parts of him lingering in me. At some point I knew: what I really needed to do was to face him, the whole complicated tragedy of him.

PART
I

Chapter 1

Mere voice is common to the brutes as man;
Articulation marks the nobler race . . .

—Alexander Bell, grandfather of Alexander Graham Bell

In 1863, at age sixteen, Alexander Graham Bell first started work on his speaking machine. He planned to give the contraption a human form, and then to play this mechanical body like an organ, with keys that depressed the different portions of the tongue and lips, and a wind chest to exhale the full words they formed. Aleck imagined that his machine would have a human skull, eyes and a nose, and a wig for hair. But first he had to build its working parts, its insides. He acquired a human skull from the local apothecary and used it to create molds for the jaw, teeth, nasal cavity, and the roof of the mouth, and he got to work, trying to coax them to speak.

In general this was a strange way for a teenager to spend his time, but in the Bell house Aleck fit right in. His father, Alexander Melville, known to his family simply as "Melville," was an elocutionist who was designing a universal phonetic alphabet, one that would be able to document any sound in any language. Melville spent much of his time poised in front of a mirror in his study, sounding a single syllable over and over again as he studied the shape of his mouth. He filled notebooks with drawings of tongue positions and assigned symbols to each sound, each tongue position and lip position, each style of breathing. Melville had grand dreams for this work, but first he had to perfect it.

Recently, Melville's research had taken him from the Bells' home in Edinburgh, Scotland, to London, with Aleck in tow. There, they visited an inventor named Charles Wheatstone. In Europe, Wheatstone was considered the father of the telegraph, which had entered the public imagination twenty years earlier. At Wheatstone's home, Aleck encountered a machine that entranced him: a wooden box with a bellows at one end and a hole at each side. There was nothing humanlike about it; its appearance was not what Aleck himself would later aspire to, but it gripped him just the same. Wheatstone threaded both hands into the box and used his right elbow to press down on the bellows, giving the machine air. Whatever Wheatstone did inside the box, Aleck didn't know; the mechanics were obscured by the box. Wheatstone operated the apparatus to utter a few sentences with its ducklike voice, and Aleck was delighted. As a parting gift, Wheatstone lent Melville the book by Baron Wolfgang von Kempelen that included the designs he'd used to build the machine. When they got back to Scotland, Melville challenged Aleck to build his own version of the machine with the help of his older brother, Melly. He wanted them to become a part of the family business, and he saw an opportunity now to stoke their interest.

Now, Aleck hunched over these parts with a coiled focus. He had a part of himself that could be free and loose, laughing loudly, doodling strange drawings in his notebooks, striking poses for his father's camera—but he tended to push this part of him aside. When he worked, he only worked. His body was a study of contradiction: black chin-length hair virtually untamable, always falling out of place; eyes full of dark intensity; posture in tight Victorian control; hands clumsier than he wished. Now he willed them to build this most delicate and powerful thing, the interior of the human mouth.

Aleck and Melly pored over the pages of the book, which included full plates detailing the workings of the human voice. Aleck learned that control of voice within the mouth begins at the back, with the tender, tissuey sponge before the roof begins: the soft palate. The soft palate has the elasticity to sink, to kiss the back of the tongue, to block air from the mouth entirely, producing nasal sounds alone, as in the beginning of the sound *ng*. The soft palate, too, needs to be able to rise for the free flow of air, a long *A. Ahhhh.* He made the palate from rubber and attached iron wire to the top, allowing him to lift the soft palate up or let it rest. Aleck knew that this precise control of physical elements was where speech was made.

Through high school, Aleck had been closer to his brother Edward, who was just one year younger, but now that he was finished with school, his work on the speaking machine would unite him and Melly in a single mission. In certain ways, Melly, who was two years older, was the opposite of Aleck. Where Aleck's default was seriousness, Melly's was playfulness and optimism. The Bells had a camera three decades before personal cameras would even begin to become common, and while other families of the 1850s and '60s stood stoically still for their portraits, the Bells donned strange costumes—plaid pants, Turkish hats, suits five sizes too big—and they played. Melly and his father were kings of exaggerated expressions: faces crushed in anguish, eyes comically wide in surprise. In one double-exposed image, Melly appears as a ghost, a sheet thrown over his body, while the rest of the family cowers playfully in horror.

By contrast, Aleck's young face was characterized by the vertical wrinkle between his eyebrows, by a look like he's squinting forever against the sunlight, lost in thought. In one photo, he has Melly's jaw yanked open, and is peering seriously into his brother's mouth, as if to see how it works.

In all things, Aleck learned through real experiences—both those that were successful and those that were traumatizing. On his best days, he learned from his failures, but normally he learned through pure enthusiasm. He always preferred open skies to classrooms, fumbling his way through his formal education. He loved to climb Corstorphine Hill behind his Edinburgh home, and wrote poetry about birds and weather, collected stones and plants and bones.

At the encouragement of his father, he had learned to classify plants by the Linnaean system, looking each plant up in a guidebook and identifying them with a long Latin name. Aleck had never received good marks in Latin, though; it was one of those subjects that drove him away from school. *Monandria*, *diandria*, *triandria*. He loved the world but hated Latin. It ruined botany. Instead, he turned to the body.

When his father gave him the corpse of a suckling pig, Aleck called for a special meeting of the "Society for the Promotion of Fine Arts Among Boys," a small club he'd started. Aleck invited the members up into the attic of 13 South Charlotte Street, which was Aleck's domain, where he collected his bones and plants and river rocks, where he laid them out in a system he supposed was scientific. The crowning piece of his collection was a human skull, another gift from his father, which kept his horrified mother at bay.

In the attic, Aleck set up boards for the young officers to sit on, and a table on which he laid the pig's body. He, the "anatomy professor," stood behind it prepared for his first lecture before his first audience, but when he brought the knife down into the abdomen, the body groaned a gassy exhale, loud enough to sound like a last gasp of life.

Aleck stood at the table, the knife in his hand, shocked. A moment later, he led the tumbling escape from the attic and down the stairs. The other boys ran until they reached their respective homes. And Aleck—no matter what coaxing, what reassurances that the pig was not alive, that he did not kill the pig—Aleck wouldn't return to the attic.

His father retrieved the corpse and disposed of it.

Despite such failures, Aleck still learned best through trials and seekings and problems, through mistakes and accidents, pounding questions that evaded every attempt at an answer, truths within truths that only experience could unearth, for better or worse.

The work on the speaking machine was coming along more haltingly than Aleck had expected. He and Melly were stuck and out of patience, but their father emphasized the importance of perseverance, of not turning away in the face of defeat. He reminded his sons of the resources at their disposal, directing them back to the book Wheatstone had lent them, to look into what it was that made voice.

As an elocutionist, Melville's work was to correct the speech of others. It was the family business: Melville's brother, David Bell, and father, Alexander Bell, were also famous elocutionists, engaged in the work of speech pathology. They worked with actors and preachers, immigrants and stutterers, to smooth out error and give power to the voice. George Bernard Shaw would draw inspiration from them for the character of Henry Higgins in *Pygmalion*, later remade as *My Fair Lady*. They helped spread the idea that not only could speech be corrected, people could also be transformed by it.

Melville saw the unique alphabet he was developing as an extension of his work as an elocutionist and also as a technological breakthrough that would be able to reach much further. Unlike alphabets before his, which largely drew their logic from the sounds of particular languages, Melville sought to shape his alphabet around any sounds that the voice was capable of making. By doing this, he believed, he could "convert the unlettered millions in all countries into readers," and pave "linguistic highways between nations."

It could allow for quick literacy in a language someone already knew, and though it couldn't teach people the meanings of words or word order in different languages, it could greatly cut down on how much time it took for people to pronounce different words correctly, as well as to read and write those words. But even without knowledge of a given language, a universal alphabet was of particular use to the British Empire in the age of missionary trips: it allowed an English-speaking missionary to read the Bible in any language needed—they would not have to know the language, only how to pronounce the words, and the Bible's teachings could reach the ears of anyone they sought to save. He understood his alphabet to be representative of something of the greatest importance: the human voice, which itself represented personhood.

The mid-nineteenth century was still ruled by the centuries-long notion that the very essence of being was embodied by speech. Voice was where language and thought met. This idea, often credited to Aristotle, had begun to meet with more biblical thinking, threading God's reflection and intention through these ancient ideas. Melville's own father wrote that "in no higher respect has man been created in the image of his Maker, than in his adaption for speech and the communication of his ideas. The Almighty fiat 'Let there be light,' was not more wonderful in its results, than the Creator's endowing the clay, which he had taken from the ground, with the faculty of speech."

Philosopher Johann Gottfried von Herder, in the late eighteenth century, argued that while all animals made sounds, it was humans who brought these sounds together into repeatable units—and in doing so, it was humans who learned to think. "The delicate organs of speech, therefore," wrote Herder, "must be considered as the rudder of reason, and speech as the heavenly spark, that gradually kindles our thoughts and senses to a flame." He believed that thought began with voice; our very humanity began with speech. To speak—and speak well—was to be able to think, to be human in the holiest, most complete form.

At the time, these beliefs were also held in their inverse: to be unable to speak was to be not quite human. Groups that did not speak, or did not speak well—immigrants, the developmentally disabled, the deaf, the poor—were often the groups who had the least access to rights. Their lack of voice had led many to deny their full humanity. In ancient Greece and Rome, infanticide of a deaf child was allowed; Jewish tradition granted deaf people neither the rights of adults nor the liability to adult punishment; Christianity often held that they could not be confirmed nor married. They were casually referred to as animals or savages.

Melville wasn't using those words, and he didn't believe in the extremes that those ideas represented. Instead, his goal was to use his alphabet to increase access to language and thus to increase access to one another. He was thinking of the many languages and cultures that were coming together under the British Empire—which within the past fifty years had slowly continued to spread to include countries as different as Yemen, Pakistan, Burma, Fiji, and Hong Kong. He was aware of the many conflicts stemming from a persistent inability to communicate. He wanted his alphabet to be a gift to the world, and though he could give lessons and lectures, he would charge nothing for the alphabet itself. He saw his alphabet as "one of the foremost Arts of Peace." He hoped the British government would cover the cost of the special typeface that was needed so he could actualize this gift.

Melville called his alphabet Visible Speech, because it acted as an instructional guide on how to shape the mouth into different sounds. Each symbol was part of a code of where to put the tongue in the mouth, how to breathe, how open the lips should be. When he was finished, Melville hoped, his alphabet would have the malleability to be used by any language, allowing anyone to participate in the power granted by proper speech.

It was because of his alphabet that he had been curious about Wheatstone's machine, though that machine echoed for him another he'd seen almost two decades earlier, in 1846: Joseph Faber's "Euphonia." Faber was an inventor who'd modeled his speaking machine after the human lungs, larynx, and mouth, placing his device within a fake torso and giving it the head of an automaton. But the effect of Euphonia's voice, which Faber operated via keyboard, was so monotonous and hollow that one viewer said that it resembled less a human than a half human, "bound to speak slowly when tormented by the unseen power outside." Faber's invention became a mockery and faded away.

But for Melville, the importance of both machines was that they showed how the human voice was, in many respects, a machine. If the voice could be replicated, then it could be controlled. If it could be controlled, it could be documented with precision and taught with precision.

It could fall into the larger movement to find a universal alphabet, a movement that had been around for over a century. All of this was central to the success of Visible Speech.

While Melville was working on stoking Aleck's interest in speech, his mother, Eliza, exerted a gentler influence, training her son's abilities in attention to sensory detail. She was a pianist who had begun to go deaf in late childhood, and her deafness had only increased with age. Now she rested the ivory mouthpiece of her hearing tube on the soundboard of their piano; she tilted her head to listen. To some extent, Eliza could still hear the instrument's resonant notes, but more so, she could feel them.

Before Melville had started his efforts to recalibrate Aleck's career path onto the family profession of elocution, Aleck had wanted to play piano. As a boy, his true love was not speech—not machines or alphabets—but music. When Eliza had taught him to play, he took to the piano full of attention and vigor. He trained his ear along the notes, the vibrations of those wires strung taut. He learned the modulations of sound, could feel and hear them in their tiniest differentiations, their dissonances and synchronicities. He could play by ear a tune he'd heard only twice.

Eliza was also an artist who sketched landscapes and ruins and rivers whenever the family went on vacation. She and Melville had met through a mutual friend, back when she was a miniature painter, painting the smallest features onto the tiniest beings. And this delicate ability was counterbalanced with strength. She was a "splendid walker," Melville said, and would walk with him for fifteen miles a day, eight days straight.

Melville had fallen in love with her quickly. She was not considered a beauty, with her strong nose, prominent chin, and gaunt cheeks, and she was ten years older than Melville besides. But Melville didn't see it that way—he thought she was thin and pretty and had "the sweetest expression I think I ever saw." At first he was filled with pity for her deafness, but he couldn't pity her once he knew her. She was well-read and widely informed. She learned the British Sign Language alphabet, but she used it only to communicate with her family. She didn't have deaf friends. He found it "philosophical" the way she saw her hearing tube as a filter, "through which nothing passed that was not worth listening to!"

As they fell for each other, they rambled through the highlands of Scotland together, bringing whiskey wherever they went, including into temperance lodgings. One day, Melville saw so many fish in the water that he decided to buy a rod and gear and try his hand at catching some. But the clear water that allowed Melville to see the fish went both ways. The fish "could *see through it* all," wrote Eliza, "and did not even require to go near his apparatus to discover the deceit." Giving up, Melville went into the water himself, as Eliza watched on, sketching.

For Eliza, sketching was both a hobby and a mode of close attention. She believed that those with poor sight or poor hearing actually observed more than others—"only, in a different way." Their thoughts were simply, she thought, "turned within," and they were more likely to keep their observations to themselves. She believed something similar was true of children: "The youth makes many observations which escape the man, merely because the latter esteems them to be not worth his notice." So Eliza taught Aleck to see with thoroughness and precision, to observe what others might ignore.

And Eliza herself was an important model. The way she lived her life flew in the face of ideas that the deaf were, at best, objects of charity, and at worst, weights on society. Instead, she taught him that a deaf person could think and observe with depth and clarity, and, too, that they could have voice. Eliza, though deaf, still spoke. She'd lost her hearing after an illness when she was eleven, long after she'd learned to speak, and so she simply continued speaking. She could rarely understand the speech of others without her hearing tube, but she could make her own voice heard. Socially this put her at a remove from congenitally deaf people, and those who lost their hearing before the age of four or so. It meant she'd had command of language and speech before her hearing began to fade. It also meant that she was able to impress upon people that she retained the ability to think in abstractions, at a time when the signing deaf were believed to only grasp the most concrete of ideas.

Being able to speak, to express language in this way, meant it was easier for Eliza to go through her days as a deaf woman. She knew almost nothing about the world of the deaf—nothing of sign language or deaf schools or deaf communities—and she slipped, with relative ease, into the world of the hearing.

Still, Aleck learned the British Sign Language alphabet to act as a makeshift interpreter when he and his mother were in public. This wasn't the same as the grammatical system of British Sign Language, some version of which had been in use for centuries already and likely beginning to standardize in the eighteenth century. Instead, this was a sign-based alphabet that could be grafted onto English. Out of the line of sight of others—below a table, or off to the side—Aleck would spell out what his mother needed to understand. When Eliza needed it most, when she really needed to grasp what had happened, or what someone had said, she could always rely on Aleck.

In public they stuck with finger spelling, but at home, Aleck would bow down near his mother's face, send his breath against her cheek. He had trained his voice to speak in a soft, deep resonance that her ears could still hear. The effort she put into these processes would have been unrelenting, but it must have seemed to Aleck then that all manner of things were possible, if only with enough patience, enough care. Eliza asked questions and, like no one else could, Aleck answered.

In the evenings, the home filled with Eliza's playing. She could close her eyes, feel the vibrations of the music on her rib cage and in the floorboards, let the sounds give what they could. Melville had once remarked that she "played the Scottish melodies with such expression that you seemed to hear the words." If Aleck could play the body with the precision that his mother played those notes, he could marry what his heart desired, to be a musician, with what his father increasingly wanted for him: a career in speech. And so Aleck began to listen for the music inside the human voice, in its perfections.

To learn about voice one must first become conscious of breathing. Before voice is anything else—before it beckons, declares, declaims, before it whispers or whimpers or cries or sings—voice is breath. The lungs fill with air, and then on the exhale the diaphragm pushes the air up and out from the lungs. Only then can a spoken word form.

Wheatstone's speaking machine offered breath, an exhale as the operator's elbow pushed down the handle of the bellows, breathing into the wind box; an inhale as a counterweight filled the bellows again with air.

Aleck had initially wanted his creation to be more human, to have a face and a nose, eyes and hair. But when these extra parts prevented Aleck and Melly from getting the basics right—the breath, the throat, the mouth—Melville talked them out of building a face. The brothers began focusing only on the mechanics. They considered Wheatstone's bellow-based lungs, but ultimately they gave up on the idea. The simplest thing was for the lungs of the machine to be their own lungs; they'd create a machine that they could blow into like a trumpet.

That settled, Melly refocused on the larynx—the voice box—its vocal cords producing volume and pitch. He made and remade it, unsatisfied. Finally he landed on a tin tube with a lid on one end and a one-inch slit through the lid, and then a piece of rubber stretched over the lid. The rubber, too, had a slit in it, aligning with the slit in the tin. When Melly blew through the mouthpiece, it made a sound, but it still wasn't right. He added another piece of rubber, staggered their slits, and produced something that he could imagine as voice. To Aleck it sounded like a toy horn, one they might blow boisterously to celebrate some holiday or another. He accepted it; he knew that the larynx wasn't really where the voice was fine-tuned.

Voice is vibration, and that vibration is shaped in the mouth. The lowest hum vibrates in the cheeks, in the teeth. Aleck still needed to make the tongue, this most complicated part of speech mechanics. Its movements seemed infinite in comparison to the quiet lever of the jaw, or the soft palate's rise and fall. He replicated the tongue with six adjacent cushioned planes of wood, each one able to move independently to create consonants and vowels. If he rigged them up to piano keys, he would be able to play this tongue: curl it to the teeth as for the letter L, flatten it to the bottom to sing a long *A*.

Eventually, Aleck and Melly created a noseless assemblage of parts functional enough to cry out a continuous vowel *ah*. Melly breathed into the machine—the air traveling through the larynx and over the tongue, against the cheeks, and releasing through the teeth as Aleck opened and closed its mouth. It created the simplest of words—*mama! mama!*—in a broken child's cry, a wounded music box.

Instead of applying their machine to anything of practical humanitarian or scientific import, the boys used it to play tricks. They took it outside to the shared stairwell of their home, and as Aleck controlled the mouth, Melly blew into it as hard as he could, and the machine whined its echoic wail.

Soon enough their downstairs neighbors creaked in quiet concern around their apartments, opening their doors and asking each other what could be the matter with that baby. Aleck and Melly squatted in their doorframe, laughing soundlessly at the spectacle, giving the creature voice again and again.

The machine stopped short of all that Aleck imagined it could do. But it was a beginning. "Our triumph and happiness were complete," he wrote, years later. Though his desire to create and control the human voice was the work that would lead him to the most troubling parts of his legacy, Aleck saw only possibility in his machine.

Melville's influence on Aleck's future was unmistakable, but it wasn't always welcome. He encouraged his son mainly in the directions of which he approved—a collection of bones, a scientific study, a speaking machine. It had been harmless enough through Aleck's boyhood, but now Aleck was sixteen, caught between looking like a boy and feeling like a man. He grew eager to escape his father's control.

At first his flights were small, just long walks to the top of nearby Corstorphine Hill where he could see the sea. Looking out at that vastness, he felt like he could do more than he was allowed. He could do more than study Latin, more than learn the function of the larynx. His life could be something of his own design—not just something his father pieced together. He began to dream of escaping, taking off to Leith under the cover of night to steal away on a ship. But in the end he turned to more practical avenues for escape, combing the newspapers for jobs in distant cities and finally securing one as a teacher at Weston House Academy, in Elgin, Morayshire.

And so he began his career as a teacher of elocution and music. He was still just sixteen, with the countenance of a man several years older than himself. He earned ten pounds a year, plus board, and his supervisor taught him Greek and Latin, helping him prepare to enter university. In his off-hours, he escaped Elgin for Covesea, a nearby seaside town with hills and caves to explore. "I spent many happy hours lying among the heather on the Scottish hills, breathing in the scenery around me with a quiet delight." He had his own money, his own time, his own plans.

His freedom made him see his own father anew. While Aleck had sometimes resisted his father's high standards and need for control, he began to soften, seeing where his father's expectations had helped him grow. Come spring, he composed a poem for Melville's birthday: "Dear Guide! Nought can thy tender care repay: / Each seeming harsh reproof was, now I see, / An act of love: received—ungratefully, / Recalling conscience forces me to say. . . ." He closed it with "Each absence makes me prize my home the more: / Return shall find me—worthier than before."

But the return wasn't as smooth as he'd hoped, his relationship with his father not as transformed as he may have dreamed—instead, it grew more tense. While Aleck had matured in his view of his father, Melville still saw Aleck as a boy. That spring, Melville began to promote Visible Speech from a more central location, relocating to his father's home in London. Aleck and Edward came along while Melly went to teach for a year at Weston House. It was here that Aleck was treated not as the teacher he'd become but as a schoolboy. He would not be asked to contribute to his father's work, only to perform it. Melville knew that in order to sell the idea of his alphabet, he would have to demonstrate its effectiveness, and so he immediately gave Aleck the task of learning Visible Speech.

Melville's script was both simple and complex. There were ten basic symbols and seemingly infinite permutations. Once a person learned the meanings of the different symbol components, they could write or read any sound—not just in language, but any sound at all. In the script, consonants took the form of horseshoes pointed up or down, left or right, depending on which part of the tongue their sound employed. The horseshoes were adorned with lines and hooks, adding new layers of complexity: use of nasality, compression of the throat. Vowels were vertical lines whose permutations of hooks and crossbars stood for the breath aperture or the shape of the lips. Glides, the sounds between consonants and vowels, had symbols. Trilling had a symbol, as did sucking. This mark signaled the direction you must blow your air, and this one indicated where your tongue should be when you do. Within five weeks, Aleck could read aloud a shutter, a wheeze, a growl, or a grunt. A year after he built his own contraption, Aleck was a speaking machine himself.

In the summer of 1864, the great phonetician, Alexander J. Ellis, traveled to the Bells' temporary home in London to test this new alphabet. Round and happy, his rumpled clothes weighted by the miscellany in his twenty-eight specially tailored pockets, he sat with Melville and worked out the most difficult and obscure sounds he could think of. Part of his thinking was not only to learn the limits and possibilities of the alphabet but also to prevent Melville's sons from guessing the words, or anticipating nuances of intonation. He came up with words in Latin, pronounced as the Estonians would, as the Italians would, as the Latins themselves theoretically would; the phrase *how odd* in any of the many accents of English; English mixed with Arabic sounds; mispronounced Spanish; and a few sounds that he made up on the spot. Melville translated them into Visible Speech.

Then he called Aleck and Edward in.

Aleck looked at the symbols, worked them out inside his mind, each word strange, often with an unexpected turn in the middle. Then he began to say them. Mostly he said them right. When he was wrong, Ellis told him so. Aleck reread the word, adjusted the inside of his mouth, and said it correctly. Or, a few times, insisted that he'd read what was written. Then Melville would examine the word and declare that he'd written what he'd heard. Then Aleck would be sent away again, Ellis would say the word again, Melville would write the word again, and Aleck, called back in, would pronounce it anew.

Even with these small errors, Ellis was taken aback by what felt like his own voice repeated back to him, over and over. Ellis, who had once made his own attempt at a universal alphabet, now praised Melville's, and imagined that, since it allowed the reader to correctly pronounce another language with ease, it could soon become "a great social and political engine."

To drum up interest and demand for the system, the Bells moved on to performing before small groups of potential funders. At the beginning of the performance, the boys would be sent out of earshot. Melville would encourage the audience to suggest any word, any sound. They offered the most impossible sounds they could think of: the sound of sawing wood, a Sanskrit cerebral *T*. One offered the sound of a long weary yawn, stretching out his arms as he did, his torso twisting. Melville translated it, and his boys returned. They studied the transcribed yawn and exhaled it, no understanding, no arms outstretched. A ghost of a yawn, one audience member described it. The audience howled in laughter to see it. But they were impressed.

Melville never revealed to his audiences what the various symbols meant and how to read them. He was still hoping to present the alphabet to the British government, or at least to raise funds to print books and spread the system that way. Secrecy would protect his investment and, potentially, his investors. So, like much in this age of medical theaters and the budding freak show, Melville's exhibitions had an air of the mysterious and mystical. Science wasn't yet separate from entertainment—experiments were often parlor tricks; science could simply be magic. By concealing the *how*, a breakthrough was a miracle. And from the first demonstration, this would be Aleck's role: He would take to the stage. He would convince people of things they believed were impossible.

Of course it didn't seem impossible to him. From his father he knew which parts of the mouth and throat controlled which sounds, how to read those sounds from an obscure alphabet, and how to give speech the power to shock, to astound. From watching his mother lean her ear toward the piano as she played, he understood how to listen very closely to the smallest shades of sound. From holding his mouth close to her ear and manipulating his own speech so that she could hear it, he began to believe that the transfer of information from his ears to her mind was a transfer of power, too. And from the way people respected her when she used her own voice, he understood how much power speech could contain and confer—not just before a captive audience, but in life.

Chapter 2

I put it to any member of the Committee whether it is not better, if his child has any peculiarities that distinguish it, that he should take every possible measure to keep them out of sight even of the child itself . . . so that when it comes into society, it shall not have anything peculiar.

—Samuel Gridley Howe

The year before Aleck started work on his speaking machine, a six-year-old girl, Mabel Hubbard, boarded a train from Boston to New York City to visit P. T. Barnum's American Museum. It was the winter of 1862, and the museum was one of the most popular attractions in New York. The predecessor to Barnum's circus, it was part exhibition hall, part theater, part freak show.

Mabel wanted to see a woman she had seen the year before: an adult no taller than she, her arms short and lean with jewels enough for every finger. Twenty-one years old, thirty-two inches tall, and twenty-nine pounds, her outfits $2,000 apiece, her pay $1,000 a week. Lavinia Warren was the stealer of hearts that year, billed as "entirely free from deformity and every drawback that would give pain to the spectator." She was, in other words, delicately positioned between what was familiar and what was unusual. Tiny and beautiful, she was the budding freak show's darling, and she was about to marry one of its biggest stars: Charles Stratton, better known as Tom Thumb.

On the train toward New York, Mabel told her mother, Gertrude, that she didn't feel well. Butterflies, Gertrude thought, and then forgot the complaint. Mabel carried with her a token from her last visit to Barnum's—a carte de visite photograph of Lavinia, no bigger than Gertrude's hand. Mabel had dubbed Lavinia "little lady" and kissed her sepia face.

It must have been such a precious thought to a child: Tom Thumb and Lavinia Warren—their wedding a little over a week away. Their love story was perfect. Charles Stratton had fallen in love with Lavinia at first sight. He wasted no time in proposing. Lavinia said she liked him but that they would need the approval of her parents. Her mother, she reminded him, disliked his mustache.

"I will cut that off," he responded, "and my ears also . . ." if it meant she would say yes.

As the train rumbled along its tracks, heat snaked through Mabel's veins, and her whole face swelled. Mabel felt the trembling wing flap of nausea as light pulsed through the car, flickering between the shade of trees and the brilliance of the sun. *Butterflies*, her mother had thought. But by the time they reached New York, there was no question that this was something more. A fever—burning, chilling. As soon as Gertrude located her parents' two black horses at the train station, she brought her daughter to the carriage.

Go. Quickly, quickly.

The fever trembled Mabel, her throat like wool, her mouth like a furnace. Her grandparents' damask curtains were drawn shut as Mabel closed her eyes to sleep it off. The rash would appear the next day, blood vessels bursting, staining her skin the color of raspberries, that first defining mark of scarlet fever. At first it seemed like the fever would pass quickly, harmlessly, but within a few days Mabel's throat swelled like it meant to strangle her.

Still, Mabel remained alert, her mind engaged. She thanked those who gave her medicine. Her grandfather wrote that she was "the most gentle docile loving little sufferer you ever saw." But the days went slow as months, each full of waiting and praying. Gertrude wrote to her husband, Gardiner, and asked him to come to New York.

Even before Mabel fell ill, Gertrude had a soft spot for the sick. A little more than ten years before, when her cook had fallen ill with Asiatic cholera, no one but Gertrude would risk caring for her. Gertrude closed the two of them off upstairs, where she would retrieve meals that were left at the top of the staircase. Finally, a proper nurse arrived and Gertrude was relieved from her caretaking, but the cook had been with Gertrude's family for years and asked for Gertrude to return. The woman died the next day, Gertrude by her side.

Gertrude could enter Mabel's room, but she was advised not to take an active role in her care. Gertrude had three other daughters to consider: an older child, also named Gertrude but whom everyone called Sister, and Mabel's younger sisters, Berta and Grace. Others were not allowed in with Mabel at all, but Gertrude came. She knew about the small coffins in which they buried children. It was how she buried her only son, fifteen years before. Now Gertrude waited with her daughter, hoping for a different outcome.

After about a week, Mabel's tongue was hard and brown. Her skin, cool. She asked her mother if Jesus still wanted little girls to pray, even if they were so sick they couldn't.

Gertrude sent for their family doctor, Dr. Putnam, who traveled down from Boston a few days later. He stayed for two nights, and then said he could do no more. There was only watching, praying, and waiting. Mabel's throat continued to swell as Gertrude watched on. Some days, Mabel would shine her blue-gray eyes open and Gertrude could believe that everything would be all right. They cleansed her in water, washing away the stains.

And then the day came when Mabel sat up, when they drew open the curtains and welcomed the light. The worst of the fever had passed, and in the days that followed, Mabel's skin sloughed in strips and scales. No one knew what the fever had done to her, what new diseases might arise in its wake to attack her weakened body. For a week Mabel lay silent, then another. When she stood and walked, it was as though she were walking in a different world; she moved with no regard to her surroundings. The fever had rotted out her inner ear, and she tottered and tumbled. She hadn't spoken a word since she emerged from the fever. She was so quiet, so distant, that the doctors believed her brain had been affected, that she would never again be the girl her family had once known.

Gertrude visited Mabel with a quiet hope. She brought Mabel her doll, her favorite books. She brought Lavinia's photograph, which Mabel took and held.

While Mabel had been bedridden, the streets around Grace Church had been mobbed for the wedding of Charles Stratton and Lavinia Warren. Inside the church, when the wedding party took to the aisle, nearly a thousand guests strained their necks, inhaled small, delighted gasps, and lifted to their tiptoes. One reporter noted that some women actually stood on their seats, "so eager in their pleasurable excitement to see, that they overlooked the possibility of being seen." At the altar, the couple walked up special small steps so that they wouldn't stumble, to a high platform before the crowd. There they promised to love each other forever. Everyone wanted to see them. Mabel had wanted to see them.

At her parents' home, Gertrude studied Mabel's face as Mabel examined the photo of Lavinia Warren. It was possible, she knew, that Mabel might not be thinking anything at all. It had been six weeks without her voice, with only her dull eyes watching. Now they rested on Lavinia. In the image, Lavinia poses beside an end table for scale, at shoulder height. She looks cool and in control, demurely lifting her skirt just enough to reveal her tiny foot, placed firmly on the ground.

Lavinia looked out at Mabel from her place at the crossroads of normal and extraordinary, and Mabel, looking back at Lavinia, said her first words in the six weeks since the fever. "Little lady," she said, and kissed her sepia face.

It was just one small phrase, but it was everything to Gertrude.

She called out, and Gardiner hurried in. Their daughter was going to be okay.

Mabel watched their familiar faces, their eyes, their smiles, their moving mouths. And then, drawing a fog over her parents' laughter, she spoke for the second time since the fever.

She said, "Why won't you talk to me?"

Aside from this question, which she asked over and over again, Mabel wouldn't speak again for months.

Mabel had deaf cousins on her mother's side, and though little is known of them and the families weren't close, Gertrude knew something of deafness. As winter turned to spring, Gardiner and Mabel made their way to the American Asylum for the Deaf and Dumb in Hartford, Connecticut. It was the first school for the deaf in America, consisting of a single brick building, in square colonial style, set back from the road by a lawn scattered with coniferous trees and cordoned off with a simple post-and-rail fence. To the Hubbards, with their opulent home in the outskirts of Boston, the school was shockingly plain.

But despite its unpretentious exterior, the school had played a central role in the relatively new national movement of deaf education. In an era when families had to go on small pilgrimages if they wanted to learn how to educate their deaf children, the American Asylum was their lodestar.

Inside its brick walls, William Turner, principal of the American Asylum, gave Gardiner the same speech he had delivered to countless other parents over his tenure at the school: for Mabel's education, they would have to wait for five years, until she was ten, and old enough to attend the American Asylum. Turner knew it wasn't ideal, but the state only funded a few years of education for each deaf child, and since almost all the students in attendance were state-supported, it seemed to make the most sense to begin that education when the child was old enough to learn practical skills.

This wasn't shocking to Gardiner, but he was worried about Mabel's voice. Before she went deaf she had some basic language; now they feared that her knowledge of language was slipping away. Gardiner had hoped the school could help Mabel retain her speech, but he suspected they had little time to act.

Turner, however, had no hope for Mabel's voice. "You can do nothing," he said.

"But she still speaks," Gardiner insisted.

Turner said that he had seen hundreds of deaf children, and Mabel was no different. She would lose her voice entirely in three months. She hadn't been speaking for long enough for her speech to be maintained and developed. "You cannot retain it," he said.

Besides, he explained, it would be easy enough for Mabel to learn to sign, to join the deaf community that was organizing around the efforts of this very school. In 1863, the American Asylum was the hub of deaf culture in America.

The typical origin story of the American deaf—that is, the deaf as a people, as a culture—begins with the American Asylum. Or, more accurately, that story begins with the origins of the school, in 1814, with a young minister, Thomas Gallaudet, and an eight-year-old deaf girl, Alice Cogswell.

Alice was playing outside her Hartford, Connecticut, home with her siblings when Thomas, her neighbor, approached her. He didn't have any formal training in education and knew she could neither hear nor speak, but he had an idea for how to try to communicate with her. He showed her his hat, and wrote *H-A-T* in the dirt. Alice responded immediately. In the months that followed, Thomas continued to work with Alice, until her father, the prominent surgeon Mason Cogswell, was so impressed that he decided to begin a school.

In the planning stages for the school, Mason Cogswell sent Thomas abroad to see how the deaf were educated in Europe. In England, Thomas met Abbé Roch-Ambroise Cucurron Sicard, director of a deaf school in Paris, who was giving a demonstration with two former pupils, Jean Massieu and Laurent Clerc, on how to communicate in signed language. Sicard invited questions from the audience, which he translated into sign language, and Massieu and Clerc answered by writing on the board. The demonstration amazed the audience—both the process and the men's literacy. The deaf, at that moment in time, were not imagined to have access to the complexities of language, and by extension, complexities of thought. However, Sicard's predecessor, Abbé Charles-Michel de l'Épée, had observed the deaf of Paris communicating with an indigenous sign language, which he brought to the school, where students quickly picked it up. As the students continued to develop language, so the language developed thought and independence in the students. Sign language was a revelation, a miracle.

Thomas followed Sicard back to Paris and began to take classes in the language of the deaf. When he had to leave, he pled with Laurent Clerc, who was thirty, to join him in America and help him begin the school. And so the deaf and the hearing formed a partnership. Together they sailed across the Atlantic aboard the *Mary-Augusta*, Laurent teaching Thomas how to sign, and Thomas tutoring Laurent in written English. Thomas answered all of Laurent's questions, about the American dinner, the American marriage, the American manners and customs. Laurent dreamed "that the people may take me for a true American citizen and not for a stranger."

In August 1816, they arrived in America, and by the fall they were traveling around New England together, giving speeches and demonstrations to raise money for the school. Laurent's abilities were the stuff of fantasy—it was hard for a hearing American audience to believe the extent of his intellect. So he framed his story in what was already known and expected: that the deaf were not full humans or citizens—at least not yet. They were, in Laurent's words, "condemned all their *life* to the most sad vegetation if nobody came to their succor, but who entrusted to our regenerative hands, will pass from the class of brutes to the class of men." That was why he and Thomas were asking for funds. He applied this same language to his own life, too. Before Laurent's education, he testified that "I had it is true a mind, but it did not think; I had a heart, but it did not feel." The framing of these thoughts might not have been helpful, but there was something real that Laurent was getting at. For deaf children, education wasn't just about traditional schooling; it was often the first time in their lives that they had unfettered access to language. At home they might have had a system of gestures that symbolized basic nouns and verbs, but it was unlikely that it was systematized into a grammar that streamlined meaning and relationships between those words, and which enables the ability to process abstract thought: emotions, ideas, even something as simple as cause and effect. Lack of language impacts everything from the ability to do math to the ability to relate to others. When Laurent argued that his education had profoundly shifted something within him, something fundamental about the way he existed in the world, part of what he was talking about was what changed when he had access to language.

After Laurent and Thomas's speeches were read or spoken, the audience was invited to ask Laurent questions. When they did, Thomas translated them into sign language, and Laurent wrote his answers on a chalkboard. The audience saw that he could, indeed, think, that his performance was a thing of truth.

After their tour, in 1817, Laurent and Thomas actualized Mason Cogswell's dream—they opened the first school for the deaf in America, the American Asylum for the Deaf and Dumb. They taught through French Sign Language, which the students picked up quickly, and which began to merge with more localized sign languages, influenced largely by indigenous sign languages, which the students brought to the school. Eventually, these together developed into what we now know as American Sign Language. Generations of deaf students would move through the school, learn to sign, become teachers, move away, and form new schools, bringing with them language, education, and lore.

Before this, deaf culture was more diffuse, less defined. There was not an agreed-upon system of values or stories through which those values were disseminated. Many deaf children, especially if they were born to hearing families, lived in isolation. It was really only the deaf children of the wealthy who received formal education. In some ways this wasn't so different from the hearing population, but the absence of education could be more destructive to deaf children, since they didn't have the mechanism to absorb certain types of knowledge from the world around them—they couldn't *overhear.*

The American Asylum offered them not only language but also community—and by extension, a stronger sense of self and possibility. By 1863, when Gardiner visited the American Asylum, deaf children were learning language and trades: cabinetry, shoe repair, sewing, and tailoring. They lived there at the school, where the teachers were often themselves deaf. Everyone communicated with each other using ASL, which in the days of Gallaudet and Clerc had been considered a beautiful, expressive language. But in Gardiner's era, it was increasingly seen through the Victorian lens as flailing, brash, overly expressive, animalistic. The fact that a people would come together around this language was considered grotesque by many in the hearing world—and the more a person's social standing depended on the Victorian ideal of restraint, the more they tended to recoil from sign language.

When Gardiner saw the deaf, he saw a people apart. Turner's words stuck with him. Inevitably, Mabel would lose her speech. Inevitably, she would be alienated from him and his world. Inevitably, she would communicate through ASL.

None of this fit into what Gardiner imagined for his daughter. It was important to him that Mabel be able to embody not just intelligence but also a certain cultural dignity. The son of a Massachusetts Supreme Court justice, Gardiner could trace his family members back to those who fought in the Battle of Bunker Hill. He had attended Dartmouth and Harvard before following his father into law, becoming a prominent patent lawyer in Boston. When Gardiner found himself inconvenienced, he was not the type of man to surrender quietly; he would rather change the world around him, and he had the means to do so. Gardiner had raced another man to organize the first street railway line outside the New York City area—and he won. He opened Cambridge's water supply system. He organized the Cambridge Gas Light Company. His home, 146 Brattle Street, sat on five acres of some of the most valuable land in Cambridge, and he counted Henry Wadsworth Longfellow among his neighbors. His family's stature was a point of pride for Gardiner, and he would not see Mabel slip away without a fight.

Now his industriousness shifted to her, to the question of how to make sure that her life didn't close in around her deafness. To him, sign language wasn't a great liberator of the deaf; it was an unacceptable marker of difference.

Sitting in Turner's office, Gardiner thought about Mabel's voice and heard Turner's words—*You cannot retain it.* Mabel absorbed none of those words, instead enthralled by the students she had seen on her way in. They were several years older than she and fluent in a language she had never seen, though she didn't yet know it was *language* that they were using. They were extraordinary, in both meanings of the word. She was taken with the grace of their hands and repelled by their difference. She believed, even then, that they would always be strange. She had no sense of the possibility of becoming like them, or that in certain ways she was already one of them.

To Gardiner and Gertrude, Mabel's voice represented the way she might be able to exist in the hearing world, the way she might come to be everything they dreamed she might become. Their dreams had not included deafness, and the idea that Mabel could speak helped them keep their dreams for her intact.

Mabel's deafness fundamentally changed the way her parents saw the future with their daughter. Deaf children seemed so *different* from them. The more Mabel seemed deaf, the more her experiences would seem inaccessible to them. And perhaps most indicative of this difference would be her adoption of an entirely new language. Potentially, their own adoption of this new language.

On the other hand if she could speak, Mabel's life could unfold, they believed, closer to how they dreamed it would unfold. Their own lives could remain relatively unchanged; they wouldn't have to embark on a path that was, to them, completely uncharted. Or at least that was the promise.

At Gardiner's meeting with Turner, Turner had mentioned rumors that in Europe, the deaf were educated to speak. He dismissed them but let slip a name: Samuel Gridley Howe, of the blind school in Boston. It came with a warning: that even if it were possible for Mabel to speak, her voice would be "so imperfect and disagreeable as to be absolutely painful." No one would want to hear her talk.

Back home, Gertrude tried to go about her days just as she had before Mabel's illness. In the mornings, she gave the dining room table a little twist this way or that to align it under the chandelier. Then, as her family gathered, she poured tea and coffee with a relaxed daintiness. Gardiner ate his oatmeal with the sugar put on last so he could see it; she picked at her own oatmeal, just a small serving of it; her youngest daughters, Berta and Grace, sat restlessly, liable to explode like "regular tinder boxes" at a moment's notice; everyone begged Sister, who was ten years Mabel's senior, to eat more; and Gertrude focused her attentions on Mabel, who was increasingly comfortable fading into the background. When breakfast was over, Gardiner would go to give his wife a kiss goodbye, and she would turn away in feigned objection.

"Oh, Gardiner," she'd say.

Their routine was the odd constant in a sea of doubt, and while Gertrude worked to maintain a sense of steadiness, Gardiner sought to find solutions, to remove the uncertainty of how to proceed with Mabel's education. By now, Gardiner had done his research, learned more about this name: *Howe.* Soon, he would be going to the blind school. Soon, he would learn what they could do.

Gardiner knew his daughter had words; he had heard them. Someone only needed to pry them from her memory, coax them from her throat. He'd followed the name *Howe* to the reports of the Massachusetts Board of Education, where he found record of a trip by Howe and Horace Mann to Europe, twenty years earlier, to study education overseas. The men traveled to Scotland, Germany, Prussia, France, and a great many countries in between. They saw it with their own eyes. In Europe, the deaf could speak.

In this Board of Education report, Gardiner and Gertrude found new stories to fill their lives, their hopes.

There was a deaf man named Habermass, known all around Berlin because he could speak. Every day strangers traveled to his home to witness it. He'd answer the door like a butler might. He would invite them in, and he'd walk them down the hall, talking the whole time, as if he were leading them to meet with Habermass. In the parlor he'd finally stop them and begin to laugh with glee. He would say, *I am Habermass!*

He was Habermass. They'd had no idea.

The report told of a teacher at the deaf school in Groningen who invited his friend to see what his students could do, how well they could talk, how closely they could read lips. His friend came to the school, and the teacher showed him around the grounds and introduced him to the students. When the visit ended, the friend turned and said, "These children are very nice, but could we visit the deaf children?"

Whether or not the stories were exaggerated, Gertrude understood the implications. She could imagine Mabel, her little girl, as a grown woman no one knew was deaf. The stories filled the spaces left in the wake of the idea of the *impossible*. And as she read the stories, two lives unfolded before Gertrude.

The first, and most obvious path, would see Mabel off to the American

Asylum. To Gertrude this seemed like a life of seclusion, in which Mabel could communicate easily with what she saw as a small group. The second path was mistier, not completely understood, not by anyone but maybe Samuel Gridley Howe. Leaders in the field of deaf education said it was absurd—but if it actually worked, Mabel would be able to speak to anyone. If it actually worked, Gertrude would still be able to speak with her daughter.

On the south edge of Boston, Gardiner stood outside Samuel Gridley Howe's school, the Perkins Institution for the Blind, and saw a school that was full of life. Porches teemed with teachers and children. He was met by a young boy, blind, who led him up the staircase to the main doors. Gardiner noticed that the boy took to the steps as easily and confidently as he himself did, no trepidation or hesitation.

At the time, the blind were described the same way as the deaf. They were "poor unfortunates," "little invalids." They were seen as incapable, in need of constant tending or care. But this school was trying to change that. The children's clothes were clean and, aside from the green ribbons wrapped around their eyes, they didn't wear uniforms. The children here, who played and studied and took part in the upkeep of the school, seemed like any other children to Gardiner. He was now even more eager to meet the man behind the school.

Howe was a man who "kindled like a torch" in the face of a great challenge. Educated as a physician at Harvard, in 1831 he took over the directorship of a new charter school for the blind—a school that grew into the Perkins Institution. Shortly after marrying Julia Ward, the suffragist and poet who later wrote "The Battle Hymn of the Republic," Howe became active in the abolitionist movement and was rumored to have offered his home to the Underground Railroad. He was now sixty-three years old, and as infused by a savior mind-set as ever, impassioned to close gaps between the privileged and the rest of the citizenship, be they the poor, women, disabled people, or formerly enslaved people recently freed by the Emancipation Proclamation. Howe knew that right could win out in the end and had clear and stubborn ideas about how to do so.

Unlike William Turner, Howe had no doubts that Mabel could be taught to speak. He told Gardiner what he'd seen in Germany, the children he'd tested himself. He told Gardiner about the resistance he faced from the American Asylum when he and Mann published their report. *Impossible*, they'd said.

But Howe, as Gardiner knew, had succeeded at the supposedly impossible before. Howe was the teacher of Laura Bridgman, who of all her senses possessed only touch. Laura Bridgman, a thin-faced girl, dainty, mousy-brown hair, fingers long and as searching as insect feelers. Before her education, she was inclined toward anger and outbursts; Howe always made sure to point that out when he told her story. She bit her teachers. Threw objects around the room. But he changed her.

Howe didn't tell the story of Laura and Asa Tenney, the hired man who had a language disorder and was known locally as an incoherent eccentric. Tenney attended to Laura, becoming closer to her than either of her parents, closer than anyone would ever be with Laura again. He took her on long walks, showed her how to feel the rush of a stream by putting a branch into it, and taught her a system of signs he most likely picked up from the Plains Indians. He taught her not to take the last egg from a hen and let her lie beside him in his bed as he read the newspaper after his chores.

"He was an Angel and a messenger to me from the hands of God," wrote Laura, who believed he had saved her life. Tenney was Laura's great anchor, her first teacher, and from the security that he embodied, she could let him lift her up in his arms, strong with work and care, and carry her from place to place, carry her body and mind through the world. There was nothing—perhaps her whole life long—that she loved more.

But after Howe came into her life, Tenney was mocked for his poverty and disability. A story of disabled people helping each other didn't have any hold, didn't make any sense, and so Tenney was all but erased from Laura's history. Howe became the person who supposedly saved her.

Howe maintained that she should learn through English, not signs; he had already invented a system of raised letters to read by. Through this precursor to braille, he taught her how to write and how to read. He created a marvel out of her: an educated DeafBlind girl. He brought her from place to place, and she performed her intelligence. Charles Dickens was beside himself when he met her.

Laura was the perfect exemplar for a growing feeling that there were whole groups of people who needed to be "restored to society," as was often said at the time. As industrialization created a desire for systems, regularity, and consistency, the word *normal* began to migrate from its former use as a mathematical term for perpendicular and toward, instead, a measure of acceptable behaviors and identities of humans. Formerly people, American Indians, the insane, deaf, blind, poor—all would have to be saved through transformation toward these new norms. Howe's school gave blind children responsibilities and trained them to go out into the world and work—to become so-called productive members of society. It was a new kind of emancipation, one that began to shift the model of lifelong institutionalization supported by charity. But that didn't mean it wasn't also problematic.

Laura, this woman who was able to move beyond the typical restrictions of her body, was the crowning achievement of Howe's career—paradoxically both because she was extraordinary and because she represented the possibilities of all. But Laura was a victory only in the abstract. As she became more self-determined, she insisted on speaking, though she could only make rough, monosyllabic sounds. She spoke loudly; Howe wanted her to quiet down.

"God gave me much voice!" she spelled into his hand.

She refused to stop speaking, and Howe told her that if she must speak, she should do it in a closed closet, where she could be as "repulsive" and "uncouth" as she wanted. Howe increasingly hid her away—he couldn't risk her disrupting her reputation as gentle and saintlike. And so she lived the majority of her life in a private room at Perkins, where she was looked upon as uncomfortably strange. She and Howe grew distant.

Still, Howe could use her story, the cleaned-up version of it, to convince people that the minds of these children were alive, waiting for answers, waiting for the world to make sense again.

Howe explained to Gardiner what he would have to do: Speak to Mabel. Don't stop speaking to her. Make sure she can see the lips. When she finally does speak, remember to reward this. Turn all attention toward her. Do not gesture. Do not point. Only speak, and only respond to speech. Nothing but speech for Mabel. She must learn that this is how she will interact.

Gardiner was the sort of man who poked and prodded his way around an idea before betraying anything of what he thought about it. His very face assisted the act: long and wise and covered with a wiry beard. Gertrude described him as an oyster: to understand anything of his own opinions, she would have to force him to open up, question by question, like a knife prying at a stubborn shell. But as stoic as his face may have been, Gardiner was filled with determination after his conversation with Howe. He was beginning to believe this whole thing was really possible, that his daughter could be saved.

Like her husband, Gertrude couldn't see Mabel as an "inmate," as the deaf students were called. She couldn't see her daughter in an *asylum* or an *institution*. She couldn't imagine her as an *unfortunate*. She said her prayer for Mabel again and again: *The Lord is my shepherd. I shall not want.* She would pray alongside her daughter daily, aware of how little she knew about what was going on in Mabel's mind, if anything.

But that same day, while Gardiner met with Howe, something shifted. As Mabel watched Gertrude recite those words, she had a flash of recognition. *Thou preparest a table before me in the presence of mine enemies; thou annointest my head with oil; my cup runneth over.*

Months of silence had passed, but Mabel remembered those movements of the mouth—and the next words, she remembered those, too. When her mother said them, Mabel joined her.

Together, they said: *And goodness and mercy shall follow me all the rest of my days.*

What it meant was everything. Not only that Mabel could speak, but that she could do all that speech represented: she could understand; she could think. It meant that she was still there, inside, and that she could be found.

Gertrude's was a prayer for acceptance, for peace, but now that was replaced with hope, with determination—with a sense of what might be. When she saw her husband's hurried steps up their walkway, she ran to meet him at the doors, throwing them open before him. She didn't need to know what Howe had said, what he believed or didn't believe. She told her husband what she now knew was possible. She told him that Mabel could speak.

Mabel's lessons began at home. If she wanted a glass of milk, she could no longer just point to the silver pitcher. If she did, her mother held it out of her reach. Any gesture or body language at all was imagined to set the whole process back. As Mabel's eyes betrayed her frustration, Gertrude repeated, "Say, 'I want a drink of milk.' Say, 'I want a drink of milk.'"

If she wanted to accompany her father for a ride in their carriage, she could no longer tug hopefully at his coattails. She had to say, *I'd like to go for a ride.*

"Say, 'Take me for a ride.'"

That was the idea, at least. But she never said these words. Instead, she closed ever inward. Gardiner wrote, "At first our daughter was very unwilling to talk."

Gertrude worked with her patiently. They worked on the twenty-third psalm, Mabel learning to read the words from the page and from the lips, just as she had when she had spoken the words before. As the months went by, Mabel remembered more and more: the songs, the hymns, the rhymes. She began to speak. Gertrude would say, "I come from dear old Boston, the land of the bean and the cod . . ."

Mabel remembered the rhythms, the shapes in her mouth. She repeated after her mother: "Where the Lowells speak only to Cabots, and Cabots speak only with God."

At the dinner table, Gardiner rarely interacted with Mabel, addressing her with single words instead of full sentences. Still, even as he shied away from engaging firsthand with his daughter and her education, he began to think that he could begin a wholly new school. He wanted to explode the preconceptions of people like Turner at the American Asylum, who had told him that speech was not possible. He got back in touch with Howe, and he began to plan.

In March 1864, he presented a petition at a hearing with the Massachusetts Legislature. It would be for a *school*, not an asylum or an institution. It would teach speech, not sign. It would admit students as young as five. It would be revolutionary.

However, members of the American Asylum were also there, and they insisted, "The instruction of the deaf by articulation was a theory of visionary enthusiasts." It would be a waste of both time and money.

The legislature acknowledged the long experience of those at the American Asylum, and largely sided with them. But the legislature also wrote that "we are satisfied the [oral] method is worth a long continued and most thorough experiment." That said, they wouldn't fund the experiment.

It was a loss for Gardiner, but not a total loss. And it gave him enough hope to continue pursuing oralism.

In the fall of 1865, three years after Mabel became deaf, the Hubbards invited an instructor named Miss Mary True into their home. She was a twenty-year-old daughter of a minister, from a small town in Maine. On her first day, awkwardly overdressed, she stood beside Gertrude and peered into the nursery at her new student.

Eight years old with long crimped hair and bright eyes, Mabel sat by the fire, cradling her doll. She looked both maternal and childlike, cooing over her little baby before she noticed anyone was watching her.

Gertrude had been Mabel's first teacher, but Miss True was something new altogether. When the Hubbards set out to find a teacher for Mabel, they didn't look for someone who had experience, necessarily—it was unlikely that a typical teacher of the deaf would work. Since the common belief was that deaf children *couldn't* speak, they needed someone without preconceptions about what the deaf could or couldn't do. Instead, they sought a certain personality: someone gentle but firm, smart but willing to take instruction.

Now Mabel watched as Miss True came into the room and lowered herself into a small chair beside her. This moment was the first of Mabel's continuous memory, the moment when Mabel's sense of her own story began: Miss True lowered herself close to Mabel's face, and Mabel watched her lips as she said, "I think I'm going to like it here with you. I think we'll do a lot of things together."

Miss True began as any teacher might, but she soon abandoned the idea of formal lessons and instead took Mabel on long walks. Through gesturing and speaking, Miss True explained the things they came upon, asking broad questions to zero in on words and concepts.

Mabel liked birch trees and their bark; she was especially good at history; kings and queens she loved; and she read voraciously, picking up new words as she went. "'Worthless' and 'Valuable' were rolled as sweet morsels under her tongue," noted Miss True. "'You are a very, very, very worthless man' settled all scores."

But pronunciation and lip-reading never ceased to be difficult. Miss True was guided only by secondhand knowledge. She knew of what teachers had done in Germany, and Gardiner helped her piece together what other teachers in America were doing. Gertrude also shared with her the things she had figured out in her early work with Mabel. But there were obvious limits to this knowledge, and she struggled to perfect Mabel's speech. In Mabel's mouth, *Cape Matapan* became *Cape Paddy Hen*; birch was *schurch*; and even *Miss True* was *Miss Rue.* Lip-reading comprehension was no better. The New Testament Bible verse "Foxes have holes, and the birds of the air have nests; but the Son of Man hath not where to lay his head," became *Oxen have horns and the birds have nests in the air, but the man hasn't anything to wear on his head.*

They sat in their home classroom, around a low table in small chairs, and Miss True improvised as best she could, teaching Mabel to say *catch* by first teaching her to say *cat-she, cat-she*, and then *cat-shing, cat-shing*, until the words *catch* and *catching*, emerged in the rush of syllables. *Which* emerged from the hurry of *whit-she*.

Mabel worked from a school reader, and Miss True never gave her reading material below her grade, believing that "she was old enough to read in the Third Reader and that was reason enough." Miss True's teaching was distinct—both in the moment, and in the speech movement to come—because she didn't much care about the miracle of speech. Decades later, Mabel would emphasize that "the mere getting of speech interested her so much less than the acquisition of knowledge." And so, while Mabel's speech remained imperfect, her intellect and independence grew.

Miss True worked with Mabel through the winter and into the spring. Meanwhile, Gardiner was working on his school. By June 1866, Gardiner had gathered money, a teacher, and five students, and opened a small experimental school in nearby Chelmsford, the first oral school in America. But Mabel didn't attend; she continued her private instruction. The following January, the legislature gathered again to hear Gardiner's plea for state support.

Teachers from the American Asylum, the pro–sign language group or "manualists," now organized more formally in their resistance. They put forward several representatives to explain their perspective. They had attempted speech education in their school for ten years before giving up on it. "Their recovery of articulation costs more than it is worth," said one teacher. The work of speaking delayed the development of intellect, and the whole idea of the connection between reason and speech was a fallacy.

Everyone agreed that the education of the deaf should ideally achieve two things: an ability to communicate with the world at large, and a stimulation of intellect. Each method—manualism and oralism—focused attention on one but was weak in the other. One manualist argued: "I think our friends who favor articulation dwell almost entirely upon this as a medium of communication with society. We regard the intellectual cultivation as the most important of the two."

They were right in that Gardiner and Howe were more focused on communication. In the years since the Civil War, normality had become an issue of national identity and security. Americans needed to be less different from each other; they needed to *cohere.* This coherence was about a set of shared national values, and the perceived threats to this coherence were many—they included the differences between the North and South, yes, but also the West, where Americans were increasingly forcing American Indian children into "schools" in order to educate them out of their culture. Threats came, too, from bodies themselves, physically and psychologically and developmentally. People whose bodies were different sometimes seemed to think differently, see the world differently. This was not seen as diversity or culture—it was seen as a threat on a unified American identity. Criminality was a threat, as was poverty. Many of these threats were seen as linked.

For Howe, the deaf were not so different from the blind in this way. He cared about education, and he cared about the mind, but he cared most about difference. "I put it to any member of the Committee," he argued, "whether it is not better, if his child has any peculiarities that distinguish it, that he should take every possible measure to keep them out of sight even of the child itself . . . so that when it comes into society, it shall not have anything peculiar."

This is what Gardiner had wanted, too, for Mabel—that neither she nor anyone else would think of her as different. Now he would use her to demonstrate what could be done. She, like Laura Bridgman before her, would appeal to people's weakness for an inspiring little deaf girl in the process of being saved.

On January 31, 1867, Mabel, now nine years old, was scheduled to appear before the legislative committee. In the room where they gathered, waiting to determine whether or not Gardiner's school should receive state support, Mabel and Miss True sat close enough for Mabel to be entertained but far enough away so that she couldn't understand the words.

For a long time, Gardiner had resisted the idea of letting his daughter speak before the legislature. He worried they would laugh at her voice, that they would think, *This is all that you've done?* But now he needed everything he could rally, and he hoped her speech would be enough. It had to be, if they were going to allow him to incorporate a new, state-funded school.

Typically, Mabel would be focused on reading people's lips, but at this hearing, as she sat and waited, she found herself transfixed by hands. There was a man at the side of the room whose arms were in constant motion: a twittering hand, a flat arm, his face wrinkled, then smoothed. Mabel could barely interrupt her gaze to ask Miss True about him. Miss True told Mabel that this was how some people talked and understood. But Mabel couldn't understand him, and Miss True said neither could she.

When it came to be her turn, Mabel took her place before the group and looked out. Like the people who talked to her sisters and not her, the men's eyes searched for someone to rest on, to see. When they finally settled, they began to ask trivial questions that Mabel found herself relieved to answer.

"Mabel Gardiner Hubbard," she said when they asked her name. She understood parts of what was going on. Her father was trying to start a school and she was here to help. "I study history, geography, mathematics." Someone would look expectant, and his lips would begin to move.

They asked questions about geography and arithmetic, and Mabel answered them eagerly. She looked over to her mother in the back, but she could only look away for a moment, because the men were still asking questions.

Someone asked, "Can you read?"

Miss True brought her book to her, and Mabel read a few pages.

Then a man leaned forward and asked, "Are you deaf?"

She blinked and wavered. Her mouth stood ajar. She looked to Miss True, who nodded, and so she said yes, though she didn't understand the question. She didn't know the word *deaf*. She didn't know what it could mean.

Chapter 3

We are satisfied . . . that there are, to say the least, forms of "dumbness," which may, by the use of Mr. Melville Bell's great discovery, be educated into some form of speech.

—*London Illustrated Times*

By late November 1865, it rained so much that the rivers of Elgin flooded. Aleck stayed up late those nights, the outside sounds muffled by the sounds of water. Aleck was training his ear, whispering his *A*'s. First the long: *ah, papa*. Then the short: *mat, gnat, cat*. He practiced until he could whisper a tune with vowels alone.

His roommate woke to find him, in the hours approaching dawn, with his face lit yellow in a mirror, a gas lamp aglow. Aleck's quiet jaw adjusted up and down, his cheeks now taut, now relaxed, his lips drawn like a purse, then loose as a lazy yawn. He snapped a finger against his throat over and over again, emitting no noise but a hollow echo.

And when his roommate asked what he was doing, Aleck would begin to explain: he was determining the scales of the sounds in the hollows, determining the pitch of the *O* or the *U*, gauging his father's correctness or incorrectness.

He was back teaching in Elgin for another year, but he was also thinking about his father's alphabet, about a conversation they'd had late one night about the scale of vowel sounds.

Melville believed that when the vowels were said in the order that

Visible Speech presented them, they had a definite ascending scale of musical tones. But Aleck was certain it was the opposite, that those same vowels in that same order created a descending scale. Melville trusted his son's musical ear, but couldn't agree on this point. In Elgin, Aleck began to look for musical notes hidden within whispered vowels.

The next night, he was back at it, with the precise control of a pencil tapping against the cheek or a thumbnail against his teeth. Aleck was no longer simply working behind his father but beside him—not only observing and learning Melville's system but building upon it. Aleck composed a forty-page letter to Melville detailing his experiments, dropping no fewer than six not-subtle hints at wanting a full set of tuning forks. He believed by using these instruments he could glean a much more specific understanding of the musical relationship between the vowels, though at this point he was limited to his meager, single C-fork. "If I could only get a box of pitch forks!!!!!"; "supposing I had a box of instruments . . ."; "Don't you think that some *valuable* results might be obtained with pitchforks???"

His father, impressed, suggested Aleck polish up his findings and send them to Alexander Ellis, the man of many pockets for whom Aleck had performed the year before. And by that winter, Aleck had a full set of tuning forks to use in his new study of sound.

But by the New Year, he also had other things to worry about.

Aleck's teaching appointment was ending at the close of this school year, and he wanted to move to Glasgow to teach and study for degree examinations at the University of London, which would confer a degree based on examination alone. It was his ticket to adulthood, to sustainable independent work.

Melville could recognize the intellectual maturity his son brought to his work, but neither Melville nor Eliza was quite ready to let their son be the adult he was. He was nineteen, but they wanted him to come home. They wanted him to study for a year at the University of London before trying for his degree. By then, the Bells had moved to Harrington Square, London, into the home of Aleck's recently deceased grandfather. Melly stayed behind to take over the work in Edinburgh.

"I can fully sympathize with your independence," wrote Melville to Aleck, "but not with the pride which would make a young man hesitate to accept with grateful alacrity the opportunity of devoting himself to study."

His mother was quick to agree. "I implore you to be *guided by your father's advice*."

The unsolicited advice didn't stop at the question of where he'd live. His mother worried about his health and advised him to stay away from pickles. His father said if he went through with Glasgow he would make himself sick cramming in his studying around a teaching schedule. His mother thought nineteen was too young to live on one's own.

"Young birds are very prone to try the strength of their wings too soon," she wrote. "The parent birds know best the proper time for independent flying."

In Elgin, Aleck had found a new haunt at the ruins of Pluscarden Abbey, its ivied walls and the sky where the roof beams once held a ceiling. On Sundays, he would escape there, lie on the grass that used to be a floor, and look up to the sky. He didn't know what to do. He loved his parents but felt trapped by their wishes and expectations. He wanted to take the university examinations and be done with them, but his parents wanted him to study and prepare. Or they simply wanted him closer to home. He was both free and tethered; both a man of his own, and a boy being called home.

Aleck chose to stay away, taking a job teaching at Somersetshire College in Bath. But he wasn't there long. The next spring, 1867, his family collapsed into mourning over the death of Aleck's younger brother, Edward, after he had struggled with tuberculosis for months. Aleck wrote in his diary, his first entry in months: "Edward died today at ten minutes to four o'clock. He was only 18 years and 8 months old. He literally 'fell asleep'—He died without consciousness and without pain while he was asleep. So may I die! AGB."

When Melville called Aleck back to London to take Edward's place as his assistant, Aleck did not protest. He must have either been exhausted of resisting, or the resistance seemed suddenly irrelevant. He finished the school year and joined his family in London that July.

By the time he arrived, Eliza had developed a weakness for sponge cakes and frequently bought Keating's Lozenges, which were meant to be taken several times daily to increase pulmonary health, staving off consumption. They were mostly morphine.

Melville gave up on trying to secure government funding for Visible Speech and revealed its secrets in his book *Visible Speech: The Science of Universal Alphabetics*. He dedicated it to Edward.

Now without any students, Aleck busied himself trying to teach the family dog, Trouve, to talk. He began by training her to growl on command and then opened and closed her lips so it sounded like she could say *ma*. *Ma* was the easiest sound to begin with. When Aleck depressed the hollow of the dog's lower jaw with his hands, she could say *ga*. Soon enough, Aleck could make her say *gamama*—a word passable as grandmama, especially from the maw of a dog. When she complied in saying *gamama,* Trouve was rewarded two treats, instead of the usual one.

Melville, meanwhile, had received a letter regarding Visible Speech. A schoolteacher, Miss Susanna Hull, contacted him to see if it might be able to help her small class of deaf children. He didn't have the time to develop the system to that end, but he was, watching Aleck coax voice from a dog. The dog's voice wasn't much, but at the same time, it was extraordinary.

Soon Aleck could form Trouve's muzzle around the sounds for *ah* and *oo*. And moving between the two in quick succession, he could shape her mouth into a sound like *ow*. By sequencing everything just right he could make her say what passed as: *How are you, Grandmama?*, which was reported with great enthusiasm by the local newspaper.

Aleck encouraged the dog to replicate this behavior without his own hands shaping her jaws, but on her own, Trouve could do nothing more than growl.

Melville didn't take on the work with Miss Hull, but he passed the job onto Aleck. The work was essentially a sort of early speech therapy, made more difficult because the students could not hear the sounds they made. Aleck didn't really think of himself as having any experience with deafness. Though his mother was deaf, he didn't imagine her as in the same class of "deaf-mutes" as those Miss Hull taught. After all, she could speak.

For him, as for most people of his day, the idea of deafness was inextricably linked to mutism. He felt that he had never seen a true deaf person, unless it was from a distance (and then, he wrote, "I kept to the other side of the road!"). He knew nothing about their education. He knew how to make a speaking machine, how to read a mysterious alphabet, how to shape mouths, but it wasn't clear to him if this would be enough.

On May 21, 1868, Aleck started his lessons with Miss Hull's students. Only two of the four students were then present: Lotty, age six, and Minna, age eight. Aleck, age twenty-one, had his father's hairline now, slightly receding; he had his grandfather's eyes, intensely serious when at rest. He began by drawing a face on the blackboard, explaining certain details through finger spelling, which he'd learned for his mother. As Aleck pointed to the drawing's lips, the point of its tongue, or the other parts of the mouth, the students touched their own. Then he erased all but what would be used to learn to speak: the lower lip, the tongue, and the upper larynx.

By then, he had the posture of a professional, his body at ease. He began to instruct the children in the same manner as his grandfather might teach an actor, or his father might teach a preacher. He taught them how to understand the body and the way it made sounds.

He taught the students the symbols of Visible Speech, whose shapes corresponded to the parts of the mouth that form speech. Soon, they were able to pronounce *p*, *t*, and *k* by recognizing their symbols. They had said these sounds before—the utterance itself was not exceptional—but the use of symbols allowed Aleck to expand these base sounds, to fine-tune Lotty's *th* to a crisper *t,* and soften Minna's hard *k* into a *ch*.

He took to this work like he took to his speaking machine, or his dog; it was an equation, a puzzle, a riddle. How to make sound perfect. What parts of the mouth, exactly, are used? How much force from the breath? How to shape the lips?

Aleck knew nothing about the debates that were beginning to unfold in America; he had no position yet on oralism. And so especially at these early stages, he wasn't opposed to using gestures to illustrate the symbols. When the children described the sound *m*, they touched their lips, clenched their hands shut, pointed to the tip of their noses to indicate nasality, and, finally, touched their throats: voice. When they put it all together they hummed an *m*.

Aleck loved this work, honing the series of small actions that led to the perfection of sounds. It was a strange skill set and one he had been raised to be particularly good at. But Miss Hull was less interested in sharpening individual sounds than she was in communication. By the fourth lesson, she requested that the children start learning words and phrases. They would go home on the eighteenth, only a few weeks away, and she wanted them to leave with words. Aleck acquiesced, though actual communication wasn't what he valued—not with these students. Instead, on his first day teaching words, he noted with quiet disappointment that "we have done scarcely anything new to-day."

By June 4, they could say *house*, *knife*, *fork*, *spoon*, and *room*, and they could form simple sentences. They still had trouble with some sounds—*laugh* became *luff* or *love*—but they pronounced perfectly *fun*, *bun*, *tongue*, and sentences such as: *Buy a dove. I love fun. How are you?*

With meaning came curiosity, enthusiasm. Even Aleck seemed to catch it, finding a playful glee along with his students. When he gave them a new set of sounds, one child would come up to Aleck with her finger up to her lips—a secret—and ask him to tell her its meaning without letting anyone else know. He would shelter his hands from the view of everyone else and somehow, either through gesture or finger spelling, explain the word he was teaching them. She would hold its meaning to herself, like the precious thing it was, for as long as she could.

The children were fascinated by *words*—what the sounds became when they learned what they meant. Aleck wrote in Visible Speech on the board; the symbols told the students how to shape their mouths and voice the sound; when they did, it became something more—it became language. There was something magical about it, that these movements of their mouths could hold meanings.

Soon Aleck's student Kate was walking around the boardinghouse saying, "I love you, Mama. I love you, Mama," practicing for the first time she would say it to her mother. She knew the words, and more: she knew what they meant.

Aleck's interest was more in mechanics, but he knew the power of language intuitively. The relationship between the two was one that his family had been working on his whole life.

That month of work would change not only the students but Aleck, too. He had thought his mother was different from deaf people—she could get around by herself, she could communicate her needs, no one really needed to know she was deaf. But now he had to wonder what that really meant. If these children could speak, maybe they wouldn't have to be deaf—not in the ways deafness was understood. They wouldn't be objects of charity or pity. They could be like his mother, able to get around just fine on their own, in need of only a little extra accommodation, and no charity at all.

Aleck was finally becoming a teacher, a gentleman, a professional in the tradition of his family. And he had found a way to apply that tradition to something new, perhaps to use it as a vehicle of social uplift for deaf children. This could be a contribution that was *his.* It was a path that he began to imagine unfolding before him, though soon he would wonder whether there was to be a path forward for him at all.

In Edinburgh, Melly continued the family work of elocution, living with his wife, Carrie, in the old family home on 13 South Charlotte Street, now with ivy climbing wildly up its walls and draughts blowing through the window seams. During the days, Melly taught students in the newly repainted classroom. When his workday was complete, he explored his newfound interest in spiritualism from a borrowed book, *From Matter to Spirit.* At night, he listened for ghosts.

In 1868, Carrie gave birth to a son. They named him Edward. For months, through the fall of 1868 and the winter of 1869, Edward was weak, and the house was unforgivingly cold.

"I have not felt so cold for a very long time," wrote Melly. "With a big fire in the room, I can scarcely hold the pen. That peculiar numb feeling is now almost gone from my feet but I still have it in my hands."

Carrie was plagued with coughs and fainting spells and nightmares, talking in her sleep, her pulse racing. Melly made mustard plaster to break up her congestion, and held it to her chest "until she moaned woefully." When she fainted, he soaked her feet in hot water, then brought her to bed, wrapped her in their warmest clothes, and then in blankets, putting hot-water bottles all around her until she sweated out the sickness and fell heavily into sleep. But the next day she'd be shivering again, coughing.

Melly missed his brother. He wrote to Aleck, begging him to write more often. "If I were not your brother and yet *felt the interest I do in certain of your hobbies*, I would be, I verily believe, overwhelmed with your correspondence." He thought maybe Aleck didn't write because Melly always teased him about his ideas—those that "strike me as ridiculous"—but he insisted that that wasn't reason enough. After all, Melly had ideas that struck Aleck as odd, too. He was drawn to spiritualism, the thought of the spirit existing in the body like a hand might inside a glove—and that the spirit could loosen from the body's grasp, could live freed from it. As he waited to hear from his brother, Melly looked for corpse lights in low-lying graveyards, where phosphorescent gases might appear over fresh graves at night. He never saw them.

As the winter wore on, Edward grew sicker. Most alarming was his breath, which began to smell like rot. "There are several distinct, sharp and well defined stinks in one," Melly wrote to Aleck, "combining in the most horrid clinging, tasteable stench that Nature's ingenuity can produce." Carrie insisted he was just teething, but Melly suspected this pestilent air might be killing them.

The house was cold, always cold, bitterly cold. Carrie had grown stronger, but soon Melly was sick. He woke with coughs spattered with red. Some days, his stomach would not settle; he vomited and vomited. "The quantity of pure bile I have got rid of is astonishing," he reported. On the days his throat bothered him, he sucked on cayenne lozenges.

Melville begged Melly to leave work behind, to take time for his health. He offered to move the family to Canada, and while at first Melly agreed, his thoughts shifted when he began to feel better. "Let me give it a fair trial before saying more about Canada," he wrote. In the end, Melly stayed in Edinburgh, learning to talk with ghosts.

Aleck didn't believe in spirits, but Melly made a pact with him anyway. They promised each other that whoever died first would try to communicate with the other from beyond the grave. At the time, Aleck had no idea about the extent of the illness spreading through 13 South Charlotte Street. He was busy teaching, studying, making a life of his own.

Melly wrote home: "Another day of fog and wintry cold. I felt very coughified all night . . . I have a beastly headache. I have always a headache in thick fog. . . . I have not been up in my workshop for a long time.

I don't know how it is, but though I have plenty of ideas, I have not got the energy to go up and work at them."

At night, Melly rubbed his chest with turpentine to keep the morning cough away. He tried to keep Carrie warm, tried to keep her resting, tried to keep her from walking through the house in only her nightgown, tempting draughts. Her voice rang out in the dark nights, speaking to her dreams. Edward's pungent breath filled the home, his illness spreading.

In February 1870, Edward died. A few months later, Aleck was called to Edinburgh to take over for Melly. When he arrived, in May, he found Melly thin, pale, weak, coughing up blood. No amount of mustard, no amount of turpentine, no amount of rest would help.

Melville wrote to Aleck, "Can it be that it is now too late? Is there no room for hope?"

Two days later, Carrie and Melly left on the 10:20 train headed to King's Cross in London. Melville wanted to bring them home. "With pillows in a cab to the station," wrote Melville, "and a bed in the train, I think he will make the journey without fatigue."

"In the meantime," he wrote, "there is life, and therefore something of a kind of hope."

But by the time Melly reached London, neither air nor water could cure him. Unlike Melville and Eliza, Carrie hadn't carried her hopes to London. Neither had Melly, not really. He knew it was the end; "death had no terrors for him," Melville wrote to Aleck on the day he died, Saturday, May 28.

Aleck received the news first by telegram, and then by letter. He had remained at 13 South Charlotte Street, the house he had grown up in. The house was empty now but for him. The classroom was empty. Melly's study was empty. The conjuring equipment Melly had borrowed from neighbors was strewn about.

In his grief, Aleck remembered his pact. During the days, Aleck began working with Melly's students; in the evenings, he played Melly's piano;

and at night, despite his skepticism, he called across that void, filled with desperation. Aleck called to his brother "in the half-hope, half-fear of receiving some communication." But he heard nothing.

Melly was gone. And Aleck's life was different now, too. On the same day as Melly's death, Melville wrote to Aleck, "Our earthly hopes have now their beginning, middle and end in you. O, be careful, and leave no opening for that fatal disease to enter." Aleck now needed to be the loyal, cautious son—as the only son left, he represented "all that [Eliza] has lost and seems in danger of again losing."

Aleck, however, was already sick. Within days, his family doctor diagnosed him with tuberculosis. The doctor gave him six months to live.

Melville summoned Aleck back to London and then built a plan to get his son out of the city, to a place where he might beat his illness. "We have resolved to give you what your constitution wants," he wrote, "a sojourn for a couple of years in the country, to tide you over a period that else might be full of danger."

But Melville wasn't talking about the English countryside. He proposed relocating the family to Canada, whose landscape he credited with his own recovery from consumption many years before. He proposed that they go for two years, that Aleck leave his professional work behind and commit to rest.

Melville made no reference to the fact that his own life would also be upended, that progress he had made in building enthusiasm for Visible Speech in England and Scotland would be lost, that his own career would be sacrificed in this effort to save the life of his last remaining son. Melville said none of this, only that the move didn't have to be permanent. That after two years, he and Eliza would follow Aleck back to England or Scotland, if that was what he wished. Only this, Melville begged of him: come with us to Canada now.

Aleck, however, was in a fog.

Days later, Aleck returned to London for Melly's burial at Highgate Cemetery, beside their brother, Edward, and their grandfather.

After the ceremony, Aleck walked the streets of London for hours, trying to see a way out of going to Canada.

These streets were deeply familiar to him. He had lived with his grandfather for several months before he began to build his speaking machine, when Aleck first began to feel his mission in life take form, when he learned to focus his mind like light through a magnifying glass, its wanderings sharpened into something so powerful it could burn.

Melly had joined Aleck on this intellectual adventure when he returned home. Melly, whose bright playful face always shined optimism where Aleck's eyes caught only shadows. "He has a stock of *good-spirits* I wish I was possessed of," Aleck had written to his mother just a few years before. Melly had once been like a partner to him, had written just months before, "I *do* wish you would give me a line now and then," and "write soon like a good chap." But Aleck hadn't written. Aleck had gone on with his work. He fought toward independence. It was everything to him.

London was where he had begun to build that independence, and it must have seemed that if he stayed he could push aside his illness, he could will everything to keep going where it was headed, move stiffly away from the past. He could pretend as though nothing was different, nothing had changed, as though there were no months of letters unwritten, as though he weren't the only family member not present for Melly's quiet death, as though there were no absences, no spirits loosened from the gloves of bodies.

But nothing was what it was before. His brothers were both dead. His own life was on the line. His father's words lingered: *O, be careful, and leave no opening for that fatal disease to enter.*

Melly had written to their father almost a year earlier: "I will treat myself as if I feared the worst—as far as care can go—and if it comes I shall at least have the consolation of not having quite lost hope."

Now Aleck looked for hope, but it was hard to say if leaving or staying offered more of it. Staying was simply to hope that the worst would not come to pass. But leaving, even if it might save his life, still felt more like dying. It was an erasure. A giving over and a giving up.

He couldn't do it. He decided not to go.

But when he went home, he found the lamplight still burning and his parents still awake. His father with his smile wrinkles creased with worry, his mother with her deep-set eyes hollowed by grief. *Our earthly hopes have now their beginning, middle and end in you.* They held hands and waited for their son's answer. Aleck found he couldn't say his decision aloud.

Instead, he said nothing, went to bed.

He was back in Edinburgh on June 2, five days after Melly's death, to begin closing out the estate. Three days later, he wrote to his father, "You know that I cannot *speak* my thoughts as I can write them." He felt his father's offer deserved a written response. "I have now no other wish then to be near you, Mama, and Carrie," he wrote, "and I put myself unreservedly into your hands to do with me whatever you think for the best." He would go to Canada.

In Edinburgh, Aleck went to work selling off his brother's old belongings: his furniture, his pictures. He garnered sixty-three pounds from it. His parents were selling off their own belongings, too.

Everything hinged on this journey. Eliza's letters were bordered in black in recognition of her state of mourning, and inside that border her words were spiked with frenetic energy. Her chest hurt; Carrie's did, too. Eliza had two dead sons, a dead grandson, and her only surviving son had been diagnosed with consumption. They had no time to lose.

Aleck dreamed of the deep forests of Canada, where he imagined he would live—some isolated, benighted place—and he dragged his feet as he cleaned out his childhood home. He wrote letters home, bemoaning the coming move, but Eliza had no more patience for it. "You don't really think you are going into the back-woods do you?" she wrote. "You are merely going into a country house, and will have civilized society there, just as much as you have here."

Aleck sold Melly's piano, and within just two weeks cured one of Melly's students of a lifelong lisp. Painfully, begrudgingly, he at last returned to London and joined his family on the ship to Canada. He wasn't angry, but he wouldn't speak. He came out only for meals. Despite the fact that his parents were bringing him there to save his life, he really didn't have any hope. He expected he was going to Canada to die.

Chapter 4

The public looked upon [deaf speech] as a sort of miracle . . . but many miracles of the past are to-day every-day facts.

—Alexander Graham Bell

When they arrived at their new home, on August 1, 1870, Canada seemed to Aleck like a clearing within a forest—no horizon, no mountains—look anywhere and there were trees, trees, trees. He had survived his journey, though no one made a record of it. We don't know whether he battled a bloody cough the whole time, or whether his illness suddenly lifted when they were out at sea, like a quiet parting of clouds. Either way, Aleck survived, and now he lived here, Canada. Just as he used to climb to the top of Corstorphine Hill, Aleck quickly went to work finding an overlook, a place where he could see beyond the trees.

He found it in a small sand cliff with a grassy depression at the top, which the family came to call the "sofa seat." Aleck, thin and pale and still recovering, would bring to this nook a pillow, a rug, and something to read. He'd sit there for hours, for full afternoons, just reading and thinking, tucked into the birches, looking out over the Grand River as it snaked off into the trees. He called it his dreaming place, which isn't to say those dreams were all happy—"there is much of sadness in the delight experienced in gazing at a beautiful landscape," he wrote years later, remembering those days. For Aleck, this was beauty and devastation, braided through with weakness and restlessness.

This Canadian forest was disappearing, Aleck could see. A few miles from where they lived, the clearings were full of stumps, abandoned. "Here and there some solitary giant tree—a relic of the primeval forest—stands mourning over the remains of its companions—to show how recently the land has been reclaimed from the wilderness," Aleck would later write.

In Canada, a friend of Melville's helped them find their home, on Tutelo Heights Road, on the banks of the Grand River in Ontario. It was situated among orchards and wilderness, and Aleck wouldn't simply stay put—not like his parents had hoped. He was soon picking apples with the orchard workers, and when he wasn't picking, he was befriending members of the local Mohawk Indian tribe, who taught him their language and war dance. Even without strength, he needed this stimulation and activity.

By the winter of 1871, Melville was starting to see his son differently. Aleck wasn't entirely well, but he was out of danger; he was still startlingly thin, but he had no cough; he was weak, but he was not lethargic. And though Aleck had agreed to come across the ocean, it was clear that he still resented Melville for the choice to move so far from home.

Aleck had no real work to throw himself into, nothing to preserve his self-respect. The nearest deaf school was 180 miles away. Aleck, cheeks gaunt, sat day after day on the sofa seat overlooking the orchard. Melville began to see his son as someone who needed a certain freedom, a certain task.

During his family's travels, Melville had stayed in touch with the educators in Massachusetts whom he'd met during a previous trip to America to promote Visible Speech. He had received a letter from Dexter King, a philanthropist who had recently opened the first oral day school in Boston, the Boston School for Deaf-Mutes. He was interested in Melville's alphabet and was looking for teachers for his school.

King's school was part of a larger wave of interest in oralism that had cropped up since Gardiner Greene Hubbard's advocacy. Four years earlier, after Mabel spoke before the Massachusetts Legislature, Gardiner won approval and launched his school, the Clarke Institution for Deaf

Mutes. Since then, other parents had begun to desire this kind of school—and they wanted it to be closer to home.

This was where King came in; he was beginning a school based on Gardiner's model. Like Clarke, it would teach deaf students to speak, but it would be different from all the other deaf schools in the region—possibly in the nation—in that it would be a *day* school.

The status quo of mid-nineteenth-century deaf education was that families who could afford it sent their children miles away from home to live at residential schools for the deaf, which were few and far between. But people were beginning to wonder about the possibility of not sending their children away at all. They were beginning to notice that deaf people who were gathered together for their educations tended to stay gathered together. Their friend groups tended to be with deaf people; their marriages, with deaf people; their news came from deaf newspapers. They were *separate.* Oralism was one way to reintegrate them, but day schools would also help ensure that the initial effect of segregation would be lessened. And since between 90 and 95 percent of deaf children were born to hearing parents, day schools meant that outside of school hours, most of these children would live in the hearing world.

The manualists, who favored ASL, had a different perspective: the time spent on oralism didn't make sense. One of oralists' primary arguments was that their method eliminated the distance between the hearing and the deaf. Under their method, the deaf could learn to communicate with whomever they wanted, so long as both people communicated in English. It was this benefit that justified the tremendous investment that was fundamental to the oralist education—at the time, Clarke was exactly twice the cost per year as the American Asylum, and students entered as young as five. However, the manualists often lived within the community where the results could be observed, and they saw that most deaf children turned their back on the oralist method as soon as their educations were over. Instead, they began to learn ASL and integrate into signing communities. One deaf teacher, Ida Montgomery, who herself had been taught to speak, wrote that now she rarely used her voice. "I speak with so much effort, mental and physical . . . and never succeed in making myself understood." In a quick poll of those around her, nine out of ten deaf people preferred sign language—even one who spoke so well as to be considered a marvel to the hearing people around her. Given these results, oralism was a significant investment with no practical outcome.

But it wasn't just that ASL appeared to be the preferred language of the deaf themselves; it was also true that deaf people had very few years in which to become educated. The schools were constantly pressed for resources, and the amount of money allocated by the New England states decreased as time went on, while the actual cost of educating increased. Since learning to speak was a painstakingly long process, oralism necessarily displaced other areas of education. Montgomery wrote: "When we consider how much our pupils have to accomplish in the very limited time given them, the question is, not 'Is articulation practicable?' but, 'Is it right to attempt to teach it?'" Teaching articulation wasn't impossible, but it meant not teaching history or math or science. It meant a dearth of deeper learning.

All of this, however, mattered little to the men spearheading the oralism movement—the *promise* of speech was enough to justify whatever sacrifices would need to be made along the way. And while they began their efforts focused on rehabilitative speech, preserving and developing existing speech skills, they were already expanding their reach, taking on congenitally deaf children and working to develop speech from scratch. Their efforts were still disparate, but they hoped the movement would soon gain in power and respect.

To make this vision a reality, King had reached out to the man who seemed to carry the greatest promise of a new, groundbreaking method, an alphabet that could transform the effort of speech. "There is an immense amount of prejudice (against the teaching of articulation) to be overcome," King wrote to Melville. King knew that his teachers would have to get it right, and quickly, before the doubts of parents and the public took over. "*We must have a success here.*"

Melville looked out and saw Aleck, restless in the trees, waiting to be needed.

He suggested to King that Aleck go to Boston and stay until the job was done. King imagined it might take as few as two months, or as many as six. It didn't matter; the pay would be the same, $500.

In addition, King would make sure the most influential educators would be aware of the work. "If your son makes a success the way will be open for him in many places," he wrote. It could launch Aleck's career in America.

Aleck left April 4, 1871, and traveled by train to Niagara Falls, where he ate, "with great gusto," a dinner of sandwiches on Goat Island. He took the 6:20 overnight train to Albany, but didn't want to spend the $1.50 for a sleeping car just to wake up again at five in the morning for his transfer. Instead, he sat upright, and for hours his head dropped down into sleep and then lurched back up again at the shock of its own fall. From Albany, he traveled to Boston, where he arrived at 3:00 p.m., and made his way to 2 Bulfinch Place.

During his time at Tutelo Heights, Aleck had never stopped longing for the city bustle, but it had been close to a year now since he had lived inside of it. Boston smelled of mud and horse and new leaves, warmed by a clean, white sun. The stones were a bright white, the bricks a good strong red. Everywhere there were horses and carriages. Here Aleck would put the past few years behind him and begin again. The whole of Boston, the whole of his new life, stretched before him.

In the next few days, he would witness the intellectual progressiveness of this city. He met with friends of his father's, who lent him books on the science of sound. He saw the public library—the first of its kind—and the new Massachusetts Institute of Technology, which opened its doors just six years earlier and admitted its first female student months before Aleck's arrival. The president of Harvard, who was a friend of Melville's from a former trip to America, would take Aleck to the State House. There, they climbed to the cupola's summit, the little room atop the dome, and looked out over the whole city. Aleck learned about the landscape and the legislature, this new country, this new world.

He also witnessed the nation's old prejudices. Within the State House, Aleck witnessed a black man testify against the Ku Klux Klan—a group that embodied the violence on which this nation's possibility was built. It was only a little over a year after the Fifteenth Amendment acknowledged a black man's right to vote, and the nation was still struggling to recover from the Civil War, which had ended only six years ago. Aleck found the man educated and well-spoken, if a little timid. He was impressed with his testimony, and appalled by the history that led to it. Later that day, Aleck saw the play *The Octoroon*, about the marriage plight of a woman who was one-eighth black. In Aleck's summary, "American Prejudice is too strong to permit of the young man marrying the Octoroon and so she dies." He left thoroughly disgusted.

Aleck didn't think this prejudice existed in his own heart, and so it was easy to dismiss it as an error of a new nation. Instead of recognizing a larger struggle between normalcy and difference, between saving and being saved, between empowerment and charity, Aleck saw only what he could dismiss himself from. And he believed that he could be a force of good here. He began to plan his classes.

At the Boston School, his lessons began April 11, at 11 Pemberton Square. At the opening of the school day, Sarah Fuller, the principal, directed the children into their seats and took her place at the head of the classroom—or the room they used for class, at least. The building they were in was not so much a school as a private residence they were using as a school. It was an upgrade from the two individual classrooms that they had been using before.

Now there were thirty-two students, all of whom clasped their hands and watched Miss Fuller's lips as she began the Lord's Prayer. She said it slowly so that even the youngest, those only six or seven years old, could follow, could learn this most necessary pattern of the lips.

Aleck admired Miss Fuller immediately: her modesty, her skill, the way she "overflowed with genuine goodness." Her face was plain and pleasant, hair parted down the middle and pulled back. In every manner she presented herself conservatively, tending toward dark unpatterned dresses with only the minimal ruffles, no bows on her neckline or curls to her hair. She gave the impression of someone serious and focused, someone much like Aleck, who was not wearing his dress coat, which he forgot in Tutelo Heights. They both seemed to understand that there were more important things than dress. More pressing things on the mind.

There was a shimmer of purpose throughout the Boston School. The teachers understood that they were experimenting with a new method, and that if they failed it would be seen as a failure not only of this school but also of the idea more broadly. If they succeeded, however, they would help to prove that speech could be taught to the deaf. More than that, they would be helping to prove something well beyond the classroom: that the deaf could be folded into the hearing world, that their difference could be erased. This was what they wanted for the deaf. The fact that this was not what the deaf themselves were asking for didn't concern them at all.

When Miss Fuller finished the opening prayer, Aleck gathered the younger students, and several of the teachers stayed to observe. He began with the face; he drew it on the blackboard. When he pointed to the parts of the face, the students pointed to their own faces. Then he began to teach the symbols of Visible Speech, which he explained to the students using gestures and makeshift signs. Within a half hour, students were able to identify and sign the names of several classes of symbols. At the back of the class, Aleck saw Dexter King, his strong face and white hair, watching on.

Later, with the older students, he built further, having them speak the symbols after signing them. Then he began fine-tuning the sounds, his whole face lighting up when a student hit on the correct pronunciation.

Before Aleck had come to the school, Miss Fuller had tried to understand the system herself, but she had come away worried that it was too complex. For the students to understand Visible Speech, they would essentially have to learn a whole new alphabet. It was hard even for her to learn it. But Aleck accomplished the task in thirty minutes. Miss Fuller wasn't the only one who was impressed—"I may say the teachers were thunderstruck," wrote Aleck.

Among them was Miss True, Mabel Hubbard's private teacher, who had taken a position at the Boston School. Now she looked on as Aleck accomplished so simply what she had spent years struggling to do. By then Mabel was traveling with her family in Europe, and Miss True had become the only teacher in Boston with any experience teaching speech to a deaf child. People looked upon that with a certain respect.

She watched Aleck closely, his alphabet, his presence at the blackboard, the white chalk outline of the face. As students said their words, he emphasized the parts of the mouth that would need to be adjusted to move closer to the correct sound. Instead of the alphabet, whose symbols are arbitrary and might represent a great number of sounds (just think of vowels), the symbols of Visible Speech related to each other. Similar sounds had similar characters, and a character stood for only a single sound. The relationships of speech were made visible, and so Aleck used his alphabet as a bridge, the visual that made progress possible and memorable.

Miss True had seniority in this setting, and since her authority diminished once this new teacher walked in the door, she couldn't help sizing him up. His work was impressive, but she was wary of the man. Aleck wasn't crazy about Miss True, either. "First impressions may be incorrect," he wrote, "but she seems to have a far higher opinion of herself than any one else has."

But the truth was, to a certain extent, that all the teachers' authority was at risk. None of them had the ability to *hear* the way Aleck did, to so quickly identify the source of imperfection in a student's utterance, to then translate it to symbols, and to write it on the board. They all watched him with astonishment, and though they were meant to learn from his techniques, it seemed impossible without his abilities. He was, they believed, *gifted.* They believed that this was what he was put on this earth to do.

Though Aleck may have been gifted, he also worked obsessively. After his first day teaching, he went back to his boardinghouse, the same boardinghouse his father had stayed in during his American visit years before. In his room, Aleck struggled with *L* sounds. He studied his own tongue in the mirror as he repeated *bottle-lifter* and *battle-laugh, battle-laugh* and *bottle-lifter*. The *L*'s all seemed different somehow. Aleck needed to iron out all the details so that he could teach. He'd worked with the students for only one day, but he was already in deep. He had identified the girl who was the likely star student, with excellent pronunciation and eventoned inflection. Her limitations would show the holes in his own understandings. How could he explain the differences? *Le, lif, laf, bottle laugh.*

He spent that evening at his desk, copying enlarged Visible Speech symbols onto pieces of cardboard. He thought about miscommunication. He had heard the story of a clergyman, just a few days earlier, teaching the gospel to the deaf. The clergyman had explained the story of Job being afflicted with boils. When he decided to check comprehension, he asked a young boy to repeat the story back to him. The boy did, with vigor, and explained that God had boiled Job for seven days.

He thought, too, of home, his mother, and of the widowed Carrie living with his parents. They had adjusted well enough to Canada, but there were still, for all of them, ghosts everywhere. Just the other day, a photo in the paper had reminded Carrie of Melly, and the whole household found itself collapsing into grief again. It may have been difficult for Aleck to be away from them, but it was also easier. He could stay busy in Boston.

The next day, April 12, was Aleck's first American exhibition. The students had worked with him for only a few hours the day before, and now they would represent what could be done. He'd planned it this way specifically to demonstrate the ease of the system and how much could be accomplished in such a short time.

He met with his students again a few minutes before ten, to review several symbols. While he did so, the school room filled with officials, parents, educators, and journalists, gathered to see what these children could do. Among them was a critical skeptic, Professor Seymour.

When it was his turn to speak, Aleck handed out copies of the Visible Speech alphabet and began to show how he taught the symbols to deaf children, demonstrating on the class gathered before them. He drew the face, isolated the parts of speech, and showed how the alphabet aligned with the different parts of the mouth.

Then, when Aleck showed the children a particular symbol, the students shaped their mouths accordingly—and spoke.

The audience marveled at the students' quick understanding, and the speed at which their words approached correct pronunciation. They looked at the copies of the Visible Speech alphabet. They sat in quiet awe as Aleck finished and opened the floor for questions.

But the whole thing looked like a sham to Seymour. He argued that the distinctions between all sounds weren't able to be captured by the alphabet: What about whispering or the differences between a sound made by voice and a sound made by breath?

Aleck was used to skepticism from his days touring with Melly and his father, but there was something more to Seymour's comments that he couldn't put his finger on. Still, he explained the importance of small distinctions in any number of languages, and the nuances of Visible Speech that were able to capture them all. But he didn't just speak to the theory. He made quick demonstrations that accompanied his arguments, swiftly winning over the room. Seymour argued that Zulu clicks were unwritable, and Aleck countered by writing them, and then, by pronouncing them.

When another man asked Seymour, at last, what he thought, Seymour said, "I am converted." But it was hard to tell what he really meant by that. Conversion on the issue was beginning to happen everywhere, but so was a larger resistance. There was a whole network of manualist schools branching out from Hartford, Connecticut, with a national network of the sort Aleck had never yet dreamed of. Overall, the country's leading schools for the deaf had seen the outcomes of oralism, and they weren't impressed. But there were many people who hadn't seen any of it, including this new method of Visible Speech. And people were hungry for the promises the oralists made.

When it came to Aleck's method, someone would always be observing, always questioning, always reporting. There were raised eyebrows, but usually they would lower again, at least at the beginning. Word began traveling, through mothers, teachers, statesmen. Word traveled to the other regional deaf schools, to other parts of the state.

"The public looked upon this as a sort of miracle . . . ," said Aleck, nearly twenty-five years later, "but many miracles of the past are to-day every-day facts." This was always how Aleck saw his work, how he would see the very idea of the miraculous. The world was always changing, the currents carrying us along, inviting ever more possibilities. It neither made sense to resist nor to stand in shocked amazement.

And yet people *were* amazed. Word traveled to New York and to Washington, DC. Soon, Aleck was in high demand. He traveled New England to teach other teachers of the deaf how to use Visible Speech.

He visited the Clarke School, where he met Gardiner Greene Hubbard. At the American Asylum, where Gardiner Hubbard had been told speech was impossible, Aleck taught a classroom of 250 deaf students the symbols of Visible Speech. Then, like a conductor, he signaled for them to speak. He arced his hand across a crowd of students, and the sound rose from their throats, loud enough to make the floor tremble, the windows rattle. The sound traveled through the halls, out the window and filled the street. The second he signaled again, it stopped. When it began again, it was louder; it was said to be heard a quarter mile away, and the pedestrians outside stalled on the sidewalks in wonder. If Aleck's technique worked as well as it seemed to, it would change everything.

Aleck believed it could. Or rather, he believed speech could, though he had always been skeptical of lip-reading, which was the other major component of the oralist education. Oralists wanted students to be able to, essentially, *hear* as well as speak. This "hearing" was to be accomplished by watching the lips of a speaker.

Aleck thought it was a sham: "My father and I thought we knew too much about the mechanism of speech to believe in such stuff as that!" He observed that many "lip-reading" students seemed to understand when their teachers said simple things like "Go to the board," "Shut the window," or "Put the book on the table," but Aleck believed it was really just the deaf child observing context and making inferences.

Out of politeness, Aleck didn't mention his doubts about lip-reading to Miss Fuller—she and the other teachers at the school seemed to put a high priority on the skill. Instead, he began to wonder if he could build a sort of speech-reading device, something to serve the same purpose of "hearing," of understanding the speech of another. He thought he could make a machine that could allow speech to be read by the eyes, a machine that could capture and transform the vibrations of voice.

Chapter 5

> Sometimes on a dark day when the clouds hang low and heavy mists obstruct the view a beam of sunlight will flash through. . . . The landscape lighting everything brilliantly and figures are seen moving about in sharp relief. Then the clouds close again and everything is blotted out.
>
> So it is with my remembrance of my first meeting with Mr. Bell.
>
> —Mabel Hubbard

By 1873, P. T. Barnum had given the speaking machine a new face. He'd bought Joseph Faber's original Euphonia—the first speaking machine that Melville set eyes on—and given it the likeness of a young woman, with dark ringlets falling around her cheeks. But the public didn't respond as he'd hoped. It had moved on from speaking machines; it wanted to gaze on people from far-off lands, bodies deformed by disease, bodies huge or small or contorted. Racial difference, ethnic difference, developmental difference, gender difference, and physical difference began to conflate under a single umbrella—*freak.* The display of the extraordinarily tall, the bearded lady, people with Down syndrome, all of this began to draw stricter parameters around what was acceptable and beautiful. And what was acceptable was a reflection of the audience themselves. In this culture increasingly infatuated with remaining "normal," there was nothing quite so satisfying as gazing at one who was not. And as anything outside of normal edged closer to "freak," efforts to normalize the deaf continued to escalate, especially from the hearing parents of deaf children. Aleck, in his office on Beacon Street, was at the center of this.

In the months since they first met, Miss True had carried in her mind that image of Aleck: the blackboard, that classroom of children learning to speak. Over the course of his work at the school she had been converted, joining those who believed he was the great hope for the deaf. The stakes for her first student, Mabel, were rising, and she thought he could be the one to help. She wasn't alone. Gardiner hadn't forgotten the man who had offered a new method to his struggling school the spring before. And he'd observed that even after a decade of tutoring, Mabel's speech was still not easily intelligible to strangers. Her interactions were mostly constrained to family. He enrolled her for lessons.

It was the fall of 1873, a year and a half since Aleck had first arrived in Boston. Mabel was fifteen, and her world was changing. Mabel's grandfather was once one of the richest men in Boston, but in the midst of this reconstruction-era depression, her family could not afford the upkeep of the home she'd grown up in. Her childhood home—with its stable, greenhouses, gardener's cottage, lawns and gardens, roses that grew up to frame the windows—now had a THIS ESTATE IS FOR SALE sign out front and dust covers over the furniture.

Now it was only Mabel and her cousins who lived there. Gertrude was in New York with her parents, and Gardiner was in Washington, DC, for work. Mabel had stayed behind in Boston just for the possibility of working with Aleck, but she ached for her family, for a life she once had. She wrote letters to her mother, begging her to come home, to open up their house one more time. "Why not say goodbye to the house where you have spent so many happy years when it is loveliest. Why not carry away a remembrance of the house as it stands in all its beauty."

In Washington, Gardiner was pouring what resources his family had into breaking up the Western Union's monopoly on the telegraph. As early as 1850, he had been looking into innovations in the field, and his work as a patent lawyer often circled around electrical inventions. He had become frustrated that telegraph offices weren't always accessible and that their rates were prohibitive for the general population—not to mention that rates varied wildly and without cause from region to region. By 1868, he had proposed a shift away from Morse's telegraph toward the Wheatstone telegraph model, designed by the man who first showed Aleck a speaking machine. Gardiner believed that with simple innovations it could carry more messages, allowing for cheaper rates and greater access to telegraphs.

He also believed that the Western Union monopoly needed to be broken if this was to happen. And he was pushing for the United States to create a "US Postal Telegraph Company," which would work similarly to, and in conjunction with, the US Postal Office. For people in the know, he was a leading authority on the subject, but he was essentially blacklisted from newspapers; the Associated Press had a business arrangement with Western Union that they couldn't afford to lose. While the cost for most ten-word telegraphs was fifty cents, newspapers paid as little as a penny a word. When one newspaper ventured to publish a story challenging the telegraph industry, their telegraph rates went up 250 percent. Another newspaper was cut off from telegraphs entirely.

For years now, Gardiner had been organizing in Washington, lobbying for a bill that would come to be known as the Hubbard Bill, for his US Postal Telegraph Company. Though his attention toward the issue had abated during the years he worked to begin the Clarke Institution, he had now turned back to it. But the lobbying was thankless and slow. In Congress, he had little power to wield, and since news traveled over telegraph wires, and AP relied on Western Union, any news coverage of his bill was skewed toward the company.

"How are you getting on with your Postal Telegraph?" wrote Mabel. "I wish so much it could all be settled up and we could all live together in some pretty little house of our own. It would be so pleasant after wandering about so long."

The family had just returned from Europe—which at the time was cheaper than living in, and maintaining, their home in Cambridge. Mabel loved the travel but hated the way it had shined a light on her deafness. Mabel spoke both English and German, but her multilingual abilities were offset by the quality of her voice. People didn't always understand, or work to understand, what Mabel said. "I am almost afraid of the many strange faces and the stares people give me when I speak," Mabel wrote. It had made her acutely aware of being deaf. She wrote that for the first time in her life, she wished she were hearing.

This wasn't just about strangers—she was approaching marrying age. She knew that her deafness could bar her from marrying someone within her class, or even from marrying at all. In the mid-nineteenth century, marriage was still typically how a woman became someone with her own home, her own respected life. Without her voice, Mabel might spend her life without a place to call her own.

She needed her lessons with Aleck to work. If she could speak better, she might marry well, and she might have a better kind of life. This she understood.

That fall, as the Boston trees started to alight in orange and yellow, Mabel and Miss True walked to Aleck's office. They passed Granary Burying Ground, which held the bodies of Samuel Adams, John Hancock, and Paul Revere, and turned into the doorway of 18 Beacon Street.

Aleck's office was up one dark flight of stairs, with walls of dingy green. Despite the floral rug, the room felt like an office being borrowed. Mabel knew Aleck's reputation was strong, but she privately considered him a "quack doctor," and his office didn't help. He was someone whom others considered capable of miracles, but she was a student who already knew that the flip side of the miracle was years of work, and years more.

None of this was assuaged by the way he looked, either. She thought he dressed carelessly, found his hair too greasy, too unkempt, and his broadcloth embarrassingly shiny: "to one accustomed to the dainty neatness of Harvard students, he seemed hardly a gentleman," she wrote.

At the time, Mabel was smitten with the famous actor Edwin Booth, with his strong, rectangular face and brooding eyes. She swooned that he was "*so* handsome, *so* graceful and withal so melancholy." Her letters to her mother were peppered with her teenage love of him. He was John Wilkes Booth's brother, but Mabel didn't hold it against him. She had followed him from afar, attending his plays as he toured through Boston. One day, she spotted him getting out of a carriage, and she reported that he was "very handsome and grand looking with long black curly hair, beautiful black eyes, and thin lips firmly closed and betraying a spirit of determination."

Aleck wasn't really the type of person she was drawn to. He didn't have the actor's swagger, the gentleman's cool control. But while Aleck didn't exude self-assurance, he was professional and confident, and he had command over his work. He understood Visible Speech, and he was one of the only people who knew how to use it to teach the deaf to speak.

Mabel already had some ability to speak, but she didn't know Aleck's alphabet, and so instead of picking up from the point of what she already knew, she now had to begin again—learning individual letters, individual sounds, sometimes only infinitesimally different from each other.

On its face, it seemed boring to Mabel, but Aleck made it somehow interesting. His body changed when he worked, took on command and control. It changed the way Mabel saw him. "He is so encouraging," Mabel wrote, "always seeming so bright and pleased . . . and he never lets me feel dull or tiresome." When he smiled, his eyes glittered with a dark excitement that she couldn't help but catch. She smiled back.

At first, Aleck was too preoccupied to think much about Mabel. He had other things to worry about. He was twenty-six years old, ten years her senior. He had just moved into the home of a woman named Mrs. Sanders, where he tutored her five-year-old deaf grandson, George, in exchange for free room and board.

He was learning American Sign Language from his adult students, so he could communicate more broadly in the community. At the time, ASL was widely considered a language that made the deaf less than human, on the level of indigenous people, or, simply, primates. A teacher of the deaf said that "like the ape, [the deaf child] is skilled at reproducing motions"; Charles Darwin wrote that signs were "used by the deaf and dumb and by savages"; Gardiner Greene Hubbard, Mabel's father, wrote that ASL "resembles the language of the North American Indian and the Hottentot of South Africa." It was considered—in either veiled or explicit ways—a rudimentary language for rudimentary beings. Earlier in the nineteenth century, it had been considered beautiful, graceful, and holy. But no longer. Now, ASL slipped into the same part of people's consciousness that categorized "freaks."

Aleck was unique in that he either didn't recognize this way of thinking or he simply didn't care. He didn't want deaf children using ASL, and his philosophy hinged on its removal from their education, but he was not so rigid in life—he knew that ASL was necessary to communicate with some of the people with whom he most wanted to converse. At this point he simply embraced the language. Later, as he came to represent the utmost threat to ASL, even his enemies admitted that, despite it all, he used the language with exceeding grace.

By now, Aleck could support himself with the money he earned from his day jobs: teaching the deaf and lecturing on speech. At night, he devised experiments: fill your ears with water to see if you can hear electricity; complete a human circuit with two people clasping hands; string a wire from this building to that. He slept when it occurred to him, usually starting around the time the milkman's jar clanked to the doorsteps.

At this point in his life, Aleck's interest in invention was intrinsically tied to the deaf. It seemed like it could hold the solution to one of his central problems: how the deaf might not only *speak* but also *hear*—or at least find some other way to perceive speech. If his deaf students could *see* their voices, they could more quickly perfect their speech. And if they could speak flawlessly, they could use English in every aspect of their lives, and then they would hardly qualify as "deaf," at least as he saw it.

His work at the Boston School was temporary, and by now he'd moved on to lecturing on deafness and speech at Boston University, and took students on the side: stammerers and stutterers and the deaf. After his lecturing, after tutoring, after dinner and piano playing, he would work on mechanisms to make speech easier to learn for his students. But he struggled, always, with his blundering hands, and with his lack of education in the relatively new field of electricity.

Even as Aleck stumbled with his speech-reading machines, he was in a better position than most teachers. The oral method was still relatively young in America. Gardiner's school had opened just six years earlier and while other schools had already adopted the method to varying degrees, the truth was that the oralist method was struggling. There was still a dearth of organized methods for teaching speech, and so teachers fumbled mostly in the dark; their students' outcomes were mixed at best.

Meanwhile, their students' general educations were advancing considerably slower than at signing schools—this no one could deny. The hope was that in the long term, the oral students would be able to catch up and then eclipse their peers. The students would have a better grasp of English; they would have made a habit of speaking; they would be able to talk easily with their hearing friends, and ultimately, to whomever they wished. But these were mere hopes, which increasingly hung from faith in Aleck's methods. It seemed that if his methods failed, oralism might not get another chance.

As Aleck's interest in electricity grew, he began to visit the recently founded Massachusetts Institute of Technology (MIT). Back then, the school consisted of only one brick building and three hundred students, but already it housed strange inventions designed to explore theories of sound.

He visited an invention called the phonautograph, which recorded patterns of sound. To use it, Aleck spoke into a stationary wooden cone, where the sound of his voice would vibrate a diaphragm. In turn, the diaphragm would move a wooden rod, at the end of which was a bristle. When spoken into, the device traced the vibrations of sound onto a pane of smoked glass, which moved past it to produce a continuous undulating line somewhat similar to a sound wave. Though it was not invented for speech training, Aleck saw potential in it. He believed that students of speech could study the patterns their own sounds created and compare them to the recorded patterns of the correct sounds.

MIT also housed the manometric capsule—another device that made sounds visible. The receiver of the capsule was similar to the phonautograph, but its vibrations were transmitted to a flame whose flickering changed depending on the sound. From the vowel sound *A*, sung on a *C* note, an agitated yellow flame rose tall from a small bud of quiet blue. An array of mirrors then reflected the agitations of fire and smoothed them into continuous bands of light. These bands of light, seen on the rotating mirrors, could be easily seen and measured. Aleck wrote of his favorite representation of sound: "A most beautiful kind of *lace-work* pattern results from [æ]. I can give you no idea of the exceeding beauty of this lovely pattern."

Aleck walked back and forth to speak vowels into each device, comparing the patterns and considering the implications. The flame was easier to read than the scrawl on glass, but as it was, it left no record. It was important, for Aleck's purposes, that the students be able to see an image of both a perfect sound, and their own sounds, so they could compare and learn from them. They wouldn't know the nuances of how to fix it, but it would at least prevent them from learning incorrect sounds. And so Aleck began to think he might use photography to capture the image of the flame-band in its many shapes, fixing these vibrations of sound in something specific and measurable. "*It is possible to teach the deaf to hear speech with their eyes*," he wrote to his mother. "I now see the key to the whole problem, and feel that I have been instrumental in making a Science of what has hitherto been an Empirical art."

As Mabel's lessons with Aleck stretched into the winter, their time often slipped away from speech and toward conversation. Aleck learned that Mabel would read for hours from her father's book collection, curled up in her home's library, with its low black walnut bookcases. Despite his role as a corrector of her speech, he was intimately accustomed to deaf voices, and loved to hear her describe her travels and the many books she'd read.

Mabel, so long confined to interactions mostly with her family and a few neighbors, was taken with Aleck, almost in spite of herself. "He was so entertaining," she wrote, "and managed to make the dullest thing so interesting with the stories of which he was brimful."

Mabel read lips so well that Aleck found he could talk to her like he could talk to anyone else—or, he would clarify, that "I could talk to her as I could *not* talk to other people!" He could talk with her about his scientific ideas, about great discoveries, about whatever he wished. He found himself growing closer to her.

All the while, she was expanding for him his notion of what the deaf were capable of. She seemed to be what he thought of as "normal," and if she was, anyone could be. She lip-read so well that he was forced to rethink his belief that the task was impossible. "Since there is in the deaf-mute no other natural defect than that of hearing," he wrote, "it is certainly possible to render him like hearing people in every other respect."

By now he was giving regular lectures around Boston on Visible Speech and its uses in teaching the deaf, making bold and impossible statements like: "there will soon be no such phenomenon as a deaf and *dumb* child."

Mabel looked forward to their every lesson, but she was also becoming worried. "I am getting along very well in my lessons . . . ," she wrote to her father, "but I have not as yet learned to apply them to conversation." She could still speak the way she always had, but she could say only a few short sentences using Aleck's methods. The weeks went by, and she guessed that her improvement was happening slowly, if at all. Aleck told her that he had never had a student who improved as quickly as she did, but it was little comfort.

Through the earliest months of 1874, Aleck continued to be preoccupied with machines and their implications for his teaching. If he could assure that the record of sound—either from the phonautograph or the manometric flame—was precise and accurate, it could help the deaf to visualize speech. But neither of them worked perfectly, so he decided he needed a device made specifically for his needs. He had no idea, at the time, how far this device would take him.

The telephone would transform the world—but it began small, as a lip-reading machine.

That spring Aleck began to devise phonautographs at home. Trying to increase the machine's precision, he changed the diameter and thickness of the diaphragm, the materials it was made of, and the shapes of its components. He prepared to present his findings in June, at the second annual convention of Articulation Teachers of the Deaf and Dumb, a convention he'd arranged.

Even as he neared the time when he would present the phonautograph publicly, he kept tinkering. The more he thought of its applications, the more he began to realize how startlingly similar the whole mechanism was to a human ear: its vibrating drum and the tiny bones that relay the eardrum's vibrations. He thought if he could better understand the human ear, he could invent a phonautograph more closely resembling this most fundamental hearing machine.

To this end, he contacted Dr. Clarence J. Blake, an aurist, who sent Aleck portions of a cadaver ear through the mail: the eardrum, the malleus, and the incus. The stapes—that tiniest, stirrup-shaped bone at the center of the ear—was missing. Aleck replaced it with a piece of hay. He softened the eardrum's membrane with glycerin and water. He constructed a cone to mimic the outer ear, and then used his device in place of the phonautograph to create an image of sound. When he sang into the cone, the sound traveled through the cadaver ear, and the hay that replaced the stylus traced a line of sound through the soot on a piece of smoked glass.

It wasn't so different from other phonautographs he'd seen, but the use of an actual human ear fascinated him. It meant that sound—mere vibrations of air—was strong enough to vibrate bone. He didn't know what this possibility could mean, or where it would lead, but it began to fill his thoughts. Was sound strong enough to be carried over distance as an electrical current?

In the mid-1870s, there were no electric power plants, no electric streetcars, no radios. Electricity was mostly used for telegraphy and fire alarms, and there was only a handful of shops throughout the country that developed electrical parts. Electrical light was being worked on by several inventors, but it was many years from being commonplace. But the telegraph *was* commonplace.

Now Aleck began to think about the telegraph. Its inner workings were easy enough to understand: at its core, it was an electric circuit being connected and disconnected in different patterns. Someone at the transmitting end sent the pattern by tapping on a key that connected the current, sending an electrical impulse through the wires—a short hold for a dot, a long hold for a dash. On the receiving end, the pattern replayed itself, mechanically connecting and disconnecting a pencil to paper that was moving beneath it, drawing out the dots and dashes. The pattern needed operators to decipher it back into alphabetic letters, but the whole system was beautifully simple. Now Aleck thought about voice, the strength of its vibrations, and its ability to travel.

Chapter 5

At the end of his days, Aleck returned home to Salem, to Mrs. Sanders and to George, the blond-haired, round-faced boy he was tutoring. Aleck walked up Essex Street to the three-story white clapboard house with green shutters, and when he could make out George's small figure, he would wave his hat in the air and George would run to the door to meet him. In his hand, George carried a glove—a talking glove, Aleck called it. It had the letters of the alphabet at different places on the fingers and the palm. Aleck had made it for George so they could communicate discreetly and with ease.

Into the evening, George would lean on Aleck's knee as Aleck spelled out his words by touching the various points on George's hand that corresponded with the letters he wished to use. Aleck told him all the adventures of the day: a monkey on Tremont Street, a dog that followed him for three blocks, a woman who carried a parrot on the train.

When George finally went to bed, Aleck would go to his third-floor study, the use of which Mrs. Sanders gave him as a twenty-seventh birthday gift. He would shut all the doors, shut himself in the solitude and silence of his work.

"When deaf-mutes are taught to speak," he wrote around this time, "people look upon it as a kind of miracle. The results which are now looked upon with such wonder will ere long become every-day facts, and future generations will look back with surprise to the time when civilized nations could allow children that were merely deaf to grow up with undeveloped minds and dumb." He saw it as his work to ensure this surprise. He wanted to make it so that the neglect of a deaf child's mind, and voice, would soon seem appalling, like the work of an uncivilized people.

When Mrs. Sanders thought he was working too hard, she would beg him to take a break, or to go to sleep. He never did. He requested his meals slipped quietly through the doorway. He was working on his phonautograph, through which the vibrations of speech might be transcribed in such a way that they could be read. In a way, it would accomplish what he currently did as a teacher: write down what the deaf student actually said so that they could compare it to the shape of the ideal set of sounds. But a device would lessen the burden on teachers. They wouldn't have to have the exquisitely tuned ear that Aleck had. And it would also allow students to practice on their own. He had written to Eliza just a few months earlier: "If we can find the definite shape due to each sound—what an assistance in teaching the deaf & dumb!!" He believed he was closer than ever to puncturing what he saw as the isolation of the deaf.

There were, of course, other solutions to this problem—isolation could be better addressed by families learning the language fully accessible to the deaf, ASL—but this wasn't what Aleck was envisioning. He continued on his phonautograph.

And as Aleck spent more and more time awake through the deadest parts of night, Mrs. Sanders took to secretly cutting his candles short, forcing him to bed as their tiny lights diminished.

Aleck didn't always meet with his deaf pupils directly, but as the Boston temperatures began to drop again, Aleck's mind drifted toward his lessons with Mabel. He wanted her to always speak the way he'd taught her, even if he were not around to guide her with his finger, which danced like a conductor's wand in the air.

Despite her initial resistance, Mabel never missed a lesson with Aleck. In the autumn's pouring rain Mabel came, and as the Boston winter settled in slow and heavy, Mabel traveled to her lessons through the snow. On February 2, 1874, despite it being three degrees below zero, despite a snowstorm and a foot of snow already on the ground, she still came to her lesson. She was the only student who came. Aleck and his assistant, Miss Locke, stood in slack-jawed awe before scurrying her in.

Mabel explained that she didn't want to miss a lesson—they cost so much, she said.

Aleck told her that he'd never charge her for a missed lesson.

But she didn't *want* to miss one.

They sat, and as the blizzard darkened the windows they lit the lamps. Tiny pellets tanged against the windowpane, and Mabel sat in front of him with her sure, shy smile. Her chin tucked back toward her body as her bright eyes drew all his attention. Outside, the whitewashed wind, the streaks melting to the glass.

With his fingers, Aleck would have outlined the diaphragm under Mabel's ribs, guiding her: Breathe from this place deep inside. Feel the sound to the core. Let the voice fill with the strength of a deep breath.

She practiced the muscles under her ribs, paid attention to her breath, pulling it and pushing it from the deepest places inside her.

Aleck loved the sound of her voice. Loved to hear it fill a room. Loved her face when she could see that he was loving her voice. Mabel sat with her ankles crossed and her torso shaped gently by whalebone. Her hands lay in her lap until the time when he reached for them to show her how a sound felt. Still, there were times when the hard static edges of sounds would break free from the shapes he had trained. That voice. All of its many beautiful sounds.

He looked at Mabel, his whole face serious. He said, "I want you to help us."

But there was something strange about the way he said it. Mabel said nothing.

Aleck, thinking maybe Mabel hadn't understood him, said it again, "I want you to help us."

"How?" asked Mabel.

He explained that he and several other teachers were getting together to issue a periodical in Visible Speech, and he was so impressed with her progress that he hoped she would contribute an original composition or translation.

Now Mabel's face was serious, too. She said she wouldn't do it. She was worried she'd make too many mistakes.

But Aleck said she wouldn't have to sign her name to the composition if she didn't want. He wouldn't tell anyone that she was the author. It would be a secret, only between them, and so she agreed.

Aleck could feel the full energy of the storm now, just outside, the groan of the walls and the glass banging in the panes. This was the type of weather that kept him up all night working. The type of weather that made him believe, if only for a moment, that anything was possible.

Soon the feeling overtook him. He accompanied Mabel home after their lesson, to ensure she got there through the storm. At the edge of a hill, they broke out running, leaping and laughing. The snowflakes, almost as large as hailstones, hit Mabel's face as they ran through knee-high powdery snow. By the time they reached the apothecary at the bottom of the hill, they were out of breath but warmed from the running. As Aleck tried to brush the snow from their coats, Mabel's veil flew wildly about them.

Back home, Cousin Mary met them at the door to size up their rumpled clothes, their winter blush. When Aleck nervously said he should be headed back to Salem, Cousin Mary agreed and promptly shut the door on him.

Alone in the whirling storm, Aleck thought of Mabel, pink-cheeked, laughing. He thought of her more and more.

Nearly a year later, in the fall of 1874, the Hubbards had all gathered back home in the old house. Aleck had become a regular visitor; it was where he was headed this October night. He was Mabel's teacher still, but now he was also friends with Gardiner and Gertrude. It must have been a sight, this raggedy Scotsman appearing regularly at one of the finest homes in Boston. "How well I remember it all," he wrote, "the blazing fire, the comfortable meal, the luxury and love . . . I looked at [the] happy home much as a friendless beggar looks into the windows of a cheerful room."

No one in the Hubbard family thought him younger than thirty-seven, a decade older than he really was. Already, his face carried weariness and work. He stayed up through the nights, working to refine his speech-reading machine. He still hadn't figured out how to use it for the deaf—it didn't actually make sound visible—but he had started to see other possible applications. A new idea had been assembling in his mind, a new kind of telegraph. It would use the same set of ideas that he was developing to help his deaf students see sound, but it would be applied to messages, not voice.

When he had mentioned it to Thomas Sanders, a wealthy businessman and father of Aleck's young student George, Thomas offered to help back the invention. A telegraph that could carry multiple messages at once wouldn't help the deaf, not directly. But if it worked it could help secure Aleck the income to keep at the work he had come to care about the most.

He must have been thinking of it that night, at the Hubbards', when he was asked to play the piano. Gertrude loved it when Aleck played. She was always asking, and he always complied. He sat down while the family prepared to listen through their ears or through their bodies.

Aleck played for the Hubbards, a sonata, but paused in the middle of it, held himself still for several beats before tapping a single note repeatedly. Then he stood over the instrument, and hummed a note to the piano. He held his finger to his lips for silence as the corresponding string echoed back the hushed note.

He looked up at Gardiner. "Did you know," he asked, "that any tuned instrument can respond to a telegraphic impulse?"

Instead of carrying the vibrations through a flame or through a needle, his idea now hinged on the theory of sympathetic vibrations: a sound created close to tuned wires would vibrate only the wires tuned to a sympathetic pitch (in this case, the same pitch). In the beginning, he saw how these vibrations could carry a sound from a speaker's mouth to a deaf person's sight. He mused that these vibrations could also carry telegraphic messages, tones in a pattern of short and long pulses.

Aleck's theory was that he could put an electromagnet under each piano string, which would initiate an electrical charge when the corresponding key was struck. He could unite the magnets in a single circuit using a wire, which would allow the charge to travel, becoming a current. That wire could theoretically be connected to another piano's strings, which would vibrate when the corresponding string on the first piano was struck. The wire could carry all the sounds, simultaneously. It didn't matter how many sounds were sent—a whole chord could be struck—the individual piano strings would respond only to the message sent at their specific tuning.

It should work the same for telegraphy. A regular telegraph wasn't pitch-specific—it was just impulses over a wire. But if each telegraphic message were tapped out, in dots and dashes, at different pitches, a single wire could carry many messages; they could be detangled at the other end.

As it was, since each wire could only transmit one message at a time in either direction, the skies of cities were crosshatched with so many wires that they nearly blocked out the sun. But Gardiner believed in a better future for the telegraph. If people could afford access to it, it would enable transmission and receipt to be separated only by moments; it could easily replace letters, bring people together, and stimulate the economy.

By now he'd devoted five years to his efforts to transform the business end of the telegraph, driving his family into debt, spending congressional sessions in Washington, lobbying for the Hubbard Bill. "The potentialities of the telegraph are boundless," he once wrote; "no man dare say what the future will bring forth." Yet whenever the future brought something forth, Western Union bought it. Gardiner was becoming a joke in Washington.

But now a gaunt Scottish teacher of the deaf stood before him, hitting a piano key, explaining pitch, vibration, wires. Gertrude was skeptical, but Gardiner was swept up. Despite his typically stoic shell, he could barely contain himself: "Don't you see," he said, "there is only one *air* and so there need be but one *wire*?"

He could see it in full: Aleck would invent the multiple telegraph, a technology by which wires could carry simultaneous messages, not just two or four, but six or eight or ten. Western Union's monopoly would be broken, making telegraphy more accessible to all. Communication would be transformed. And, with the help of this strange young teacher, he would be at the center of it.

Chapter 6

> I am like a man in a fog who is sure of his latitude and longitude. I know that I am close to the land for which I am bound and when the fog lifts I shall see it right before me.
>
> —Alexander Graham Bell

Early in the winter of 1874, at Charles Williams's electrical shop on 109 Court Street, Thomas Watson was working on a torpedo exploding apparatus, one that could be used to explode naval mines. He was crouched over his table in tight concentration when a customer burst into the workroom. Tall and quick and concerned, black hair in bushy disarray, mustache drooping, he carried two electrical devices in his hands as he headed straight for Thomas.

Thomas had no idea who the stranger was or what he was doing in the workroom, but when he recognized the devices in his hands he began to piece it together. The man was holding two electromagnets, each with a flat piece of steel attached, and a clock spring mounted over that. Thomas, a young apprentice in the shop, had built them. Which meant that this man, Aleck, must be the inventor.

In Williams's shop, Thomas Watson most often worked at a hand lathe turning binding posts, small electric connectors. For five dollars a week, he turned the lathe and felt grateful that he no longer had to work clerking, bookkeeping, or carpentry jobs. This was one of the largest electrical shops in the nation, with between thirty and forty men working at any one time, but the electrical industry itself was still crude and modest, and the equipment reflected that. The men worked with their hand lathes or with vises and files, aspiring to one day work the single sixteen-inch engine lathe, their prized piece of machinery. No one worried too much that its middle wiggled badly. They found ways to compensate.

The apprentices flew about the shop like the hot chips that flew from the lathe. Nimble with their fingers, they made call bells, annunciators, galvanometers, telegraph keys, telegraph printing parts, sounders, relays, registers. They did the regular work of the place, which wasn't invention—it wasn't an invention shop. They worked on things other people had already invented, assembling electrical objects and crafting the parts that made them tick.

Inventors, Thomas had come to understand, barely knew how anything worked. Heads full of ideas, pockets empty. They came to Williams's shop with ideas, and they liked to work with Thomas, who had both patience and prowess. Thomas liked to work with them, too, because even though things almost never worked out the way the inventors imagined, he liked to live in those unexpected, unpredictable worlds. He had bought himself a copy of *Davis's Manual of Magnetism* for twenty-five cents, and studied it at home when his workday was done. In his journal, he jotted down his own questions and ideas about the world:

"Is red hot glass a conductor of electricity?

"What sort of a vibration is a hissing sound?

"Device to light kerosene lamps by electricity.

"Why do clouds that are heavier than air remain suspended? Is it an effect of electricity? . . ."

Thomas was on the brink of twenty years old, but his face still held on to its boyishness, with round features and muttonchops that sat on his cheeks as if to prove he could grow them. He looked full of a dreamy hopefulness, though he had a hard-nosed upbringing. His father was a foreman at a stable, working from early morning until late at night, seven days a week. His mother and half sister kept house—washing, ironing, scrubbing, mending, cooking—a life Thomas saw as full of drudgery. "There were no conveniences to lighten their labor," wrote Thomas. In Williams's shop, he had settled on work that would help create those conveniences.

Now Aleck held the two small devices out to Thomas, whom he'd deduced was the builder, and began not by complaining or chastising, but by explaining. He wanted Thomas to understand his idea as a whole, wanted him to understand what he was building. In the middle of the shop, to the flummoxed young apprentice, Aleck laid out the principles behind his "harmonic telegraph."

It was an expansion of the idea he'd proposed to Gardiner just a few months earlier, that the motion of one vibration would send into motion an echo of it on a nearby piano string of corresponding pitch. This wasn't the telephone that he would become known for, but an earlier step in that journey: an improvement of the telegraph.

Aleck would create an electrical circuit, a path for an electric current between a transmitter and a receiver, both of which would include flat pieces of metal tuned to the same frequency. The transmitter would initiate the current using a battery or magnets, and the pieces of metal worked like those piano strings Aleck had first used to explain his theory to Gardiner. They vibrated when they were in proximity to a sound that was the same frequency as their tuning. The various frequencies would travel over a single wire, but because of these pieces of tuned metal, the messages would theoretically be sorted out again at the receiving end, somewhat like sorting a piece of music back out into its individual notes. Aleck could theoretically send as many messages as he had receivers tuned to different frequencies. His plan was for six, he explained to Thomas. If it worked, he could send six telegraphic messages along a single wire.

Only a few weeks before Aleck burst into the workroom, Gardiner Greene Hubbard had joined with Thomas Sanders to fund this invention. Gardiner's patent-attorney background and business savvy were significant assets—or at least Aleck must have thought so. Thomas Sanders was less thrilled at the new addition, worrying vaguely that Gardiner was corrupt, that he was in it only to line his pockets. Still, in the end, they all struck a deal: Aleck would work on the invention, a telegraph to carry multiple messages simultaneously; Gardiner and Thomas Sanders would support the costs of his invention, but not his living expenses or his time. In the end they would share the profits. Aleck now had the ability to pursue work that might help him gain financial stability, though he still understood his first obligation to be to the deaf.

He spent his nights and weekends tinkering away at his invention. He researched in the Boston Public Library, which had opened only nineteen years earlier as the first large, free public library in America. The doorways and ceilings were opulent, but many of the bookshelves were still empty. There, he read up on electricity and acoustics.

Before he even got off the ground, though, news reached him that a well-established Chicago inventor, Elisha Gray, was now able to telegraph music over twenty-four hundred miles using a method and theory shockingly like Aleck's. In a *Chicago Tribune* article, Gray explained that if one took a vibrating tuning fork and placed it near an organ pipe of the same key, the pipe will begin to vibrate in unison. "I believe," said Gray, "that the principle is applicable in sound-telegraphy." He suggested that at least twenty messages could be sent at once, though he had only succeeded in sending four.

The news was alarming. Aleck didn't yet care about telegraphy for telegraphy's sake, but Gardiner immediately began applying pressure on Aleck to move more quickly. He believed that Aleck's invention could be saved. Elisha didn't display a complex understanding of the invention, nor its important parts, Gardiner argued. He encouraged Aleck to keep detailed notes of everything he did. "You must not neglect an instant in your work," he wrote.

Aleck was worried about Gray, but not too worried. He wrote to his parents: "He has the advantage over me in being a practical electrician—but I have reason to believe that I am better acquainted with the phenomenon of sound than he is."

His mother wrote back: "Work as if you were certain of success, in whatever you undertake, but always with the reservation of it may be, not that it will be." And so, in the spirit of working as though certain of success, Aleck entered into his collaboration with Thomas Watson.

By January 1875, within several weeks of their first meeting, Aleck and Thomas were working together regularly in Williams's shop. Aleck was determined to not only theorize but also to make the invention *work*. He ordered six pairs of transmitters and receivers so that he and Thomas could try out a fuller harmonic array. But in the end, they didn't work out. The receivers picked up the wrong messages.

The disappointment was acute for Aleck, especially since Gardiner had made clear a need for haste. But Thomas, who worked regularly with struggling inventors, was used to things taking longer than expected. Tinkering was part of the process, and it couldn't be rushed. Besides, he admired Aleck, and they had both come to enjoy the time they were spending together. Aleck was the first educated man Thomas had known, and Aleck began bringing books to his new friend: Tyndall and Helmholtz and Huxley. When Aleck found that Thomas had abandoned his studies of algebra, he bought him the newest algebra book he could find. At Mrs. Sanders's home, Thomas first witnessed the use of forks—he had always eaten by knife. And unlike the straight, two-tined forks that Thomas had seen before, Mrs. Sanders's were curved like a spoon and had four tines. Thomas carefully observed the use of "this leaky, inefficient implement," and practiced on his own until he could use it with ease.

More than anything, though, Thomas marveled at Aleck's ability to play the piano, which had always seemed to Thomas to be the "peak of human accomplishment." He studied Aleck's fingers as they made music, trying to understand the mechanism. He didn't know if each key needed to be struck precisely, or if the pianist needed to just strike any key in the general vicinity—"it seemed so occult and inexplicable."

Thomas, too, lived in Salem, and he and Aleck would sometimes go for Sunday walks together. Aleck, with his Scottish accent, talked about many things: piano, speech, flight. Thomas wanted to learn how to speak the way Aleck did: "The tones of his voice seemed vividly to color his words. His clear, crisp articulation delighted me and made other men's speech seem uncouth."

As they walked along the beaches of Swampscott, with the salty smell of sea and kelp, Aleck gave Thomas informal lessons, and Thomas recited poems. *It was many and many a year ago, / in a kingdom by the Sea, / that a Maiden there live, whom you may know / by the name of Annabel Lee . . .*

On one walk, as they came upon the body of a dead gull, Aleck stopped. He bent to examine it, spread its wings, releasing a smell like petrichor and evaporating sea, brine and musk, long-wet feathers. He measured the wingspan and then lifted its body to estimate its weight. Thomas didn't get too close. He watched from a distance as Aleck admired the bird's mechanics: the lines, the functions of the wings, the bones he knew to be hollow.

Aleck spread the gull's graceful wings and continued explaining to Thomas one of his most spectacular, most ludicrous ideas: Aleck believed he could make a machine that could fly, that could lift itself up into the air and hold things heavy, could hold people, could defy physics, about which he knew admittedly little. At the time, the Wright brothers were ages ten and four. The world was nearly three decades away from the first airplane's flight.

Thomas listened patiently to this idea as he listened to all of Aleck's ideas, with a bit of measured skepticism melted by Aleck's firm enthusiasm. There was something about Aleck, the way he scoffed at impossibility, the way his mind wasn't limited by what other people said couldn't be done. He wasn't yet an American citizen, but already there was a distinctly American ring to his boldness and his naiveté. He didn't let other people tell him what couldn't be done, what shouldn't be done.

Thomas was more measured, but at Williams's shop, he had learned to doubt less and believe in minds greater than his. He was always hoping to be surprised, for his mind to stretch around the possibility of a new idea. Thomas moved upwind of the dead bird, watched Aleck, and listened. Aleck's ideas—ideas that friends would warn him against uttering lest he sound mad—filled Thomas's mind even when he wasn't working on Aleck's contraptions. They filled his world with a sense of magic.

That winter, Thomas and Aleck took over the attic of Williams's shop, the walls unglazed red bricks, the steeply slanting ceiling exposing the wooden roof beams. A brass gas lamp hung from the ceiling; another, with its emerald glass shade, sat on the long workbench that ran the length of the wall.

They strung wires all about: from the pegs in the walls, along the beams and the bench. There were batteries and magnets that they could use to initiate electric charges to power their devices. They created imaginary New England locations at different places in the garret, but when they sent messages from Portland, in one corner, to Salem, in another, only about half the message would arrive—the rest of it scattered across the other stations. Their tuned receivers were meant to pick up only the message at their frequency, but in the end, Aleck and Thomas couldn't seem to keep their dots and dashes together; the words dissolved to letters, to beeps, some here and some there.

They could successfully transmit two or three messages along the wires at once, but they would need to get that number up to at least four in order to have any commercial value. Aleck and Thomas tried again and again. Thomas tapped out their messages, and Aleck waited to receive them—but again and again, Aleck's dark face betrayed only blankness or frustration. The tests, according to Thomas, were to no productive end: "The chief result attained [was] to prove to Bell that the harmonic telegraph was not as simple as it seemed."

Aleck did what he always did when things weren't going right: he began to dream of something new. He was thinking again about voice, about a device he'd begun to envision of the summer before, another receiver that could respond to a wide range of tunings simultaneously. Like his multiple telegraph, it relied on the ability to transmit various pitches across distance—but unlike the telegraph, it would serve not to transmit several different Morse code messages at each pitch, but instead to utilize all the pitches to transmit a single message, one that could carry voice. It was the fundamental idea of the telephone. But, like his flying contraption, it was only a dream.

By late January 1875, Aleck wrote to Gardiner that he had working models of his multiple telegraph ready—"(though I am not yet perfectly satisfied with all the instruments)." He may not have been satisfied, but he didn't have time to work on them any longer. He needed to return to his teaching work, and he feared they were running out of time to compete with Gray. He thought they should file a patent for the models as they were.

They were, in fact, good enough. By February 1875, as Aleck and Gardiner finalized their patents, Aleck and Thomas continued their tests. They could transmit two messages at a time over the same wire, sending the words *Multiple Telegraph* at the same time as they sent *Autograph Telegraph*. Then, still over a single wire, the words *Katie Silsbee*, *Valentine*, and *Well Done* at the same time as *Multiple*, *Cat*, *Lizzie*, and *Lovely*. Other times, they set up four stations on a single circuit, sending different messages from, and to, different stations. Gray was on the forefront of their minds, and the lawyers Gardiner had hired strategized about how to file Aleck's patents to interfere with Gray's.

Gray had already filed a number of patents for similar inventions, but that didn't necessarily mean that Aleck's patents wouldn't win out in the end; patent law wasn't about who filed first, but who *invented* first. By writing patents in a certain way, the lawyers could initiate a formal hearing, called an interference, to determine who came to the invention first.

Two nights before Aleck was to leave for the national patent office in Washington, DC, he tested the telegraph one last time. At first everything was fine, but one by one, the transmitters stopped transmitting. Some lasted only five minutes, others lasted thirty, but they all stopped working. "*I felt that I could not trust them in Washington*," wrote Aleck to his parents. He had planned to conduct demonstrations of the telegraph during his travels, and so it had to work perfectly. He and Thomas worked all the next day and into the night, getting them back in working order. By February 13, Aleck left for the capital.

Between February 25 and March 10, he and his lawyers filed three separate patent applications to cover the different concepts behind the harmonic telegraph. While he was in DC, Aleck also met with Joseph Henry of the Smithsonian Institution, and shared with him not only his telegraph, but his most precious idea: the telephone. Henry loved it, calling the idea "the germ of a great invention." Henry may have been the first person in Aleck's life to really support the idea. And if anyone would know an idea worth pursuing, Henry would.

"My visit to the Smithsonian Institute seems to me to be the brightest spot in my whole life," he wrote to his parents. It meant he was accepted into the scientific community. That acceptance, though, had an inverse meaning, too. He was accepted, yes, and now he was a threat to other inventors.

His first patents were officially filed, but he still hadn't heard whether they were accepted. In the meantime, Aleck traveled to New York City to test his instruments for Western Union in hopes that they might go into business together. Their head electrician inspected Aleck's telegraph, asked questions about its mechanics and inception. Finally, he asked Aleck to demonstrate the machine.

"The signals were as clear, sharp and rapid as with the ordinary Morse sounder," wrote Aleck. The company's electrician was so impressed that he wanted to experiment with them on an actual telegraph wire—one that ran from New York to Philadelphia. Aleck's model was not intended for that, though; the magnets and batteries were designed for trials over shorter distances. No one had high hopes for the effectiveness of the trial, and yet, wrote Aleck, "they *did work.*" They weren't strong, but they were sharp and clear over two hundred miles of wire.

That afternoon, Aleck met with Western Union's president, William Orton, who put his feet up on a chair and asked Aleck about the invention—how it worked, how it began, the dates of different advancements. After Aleck had described the invention in depth, Orton told Aleck that Elisha Gray had been to see him earlier that day, that he was very impressed with Gray's work. He added that Aleck's work was crude compared to Gray's, and then, casually, delivered the final blow. "Is Mr. Gardiner Greene Hubbard interested in this matter with you[?]" he asked. When Aleck confirmed, Orton said he'd never support a venture that would benefit Hubbard, after all that Hubbard had done to attack Western Union. Now that he had everything he needed from Aleck, Orton drove Aleck back to his hotel.

"You seem to think my anxieties are over," he wrote to his parents upon returning from his trips, "when in truth they are really only beginning." He was rushing not only to make the telegraph work reliably but also to take out foreign patents and to commercialize his invention before Gray swept in. On top of everything, he still didn't know if he would be granted the patents he filed, or if they would go to Gray after all. His anxiety mounted until he was quite ill, his doctor begging him to give up as much work as he could.

From Canada, his family grew worried. Eliza wrote, "The worst of it is, that even if you gain the [multiple telegraph], that restless brain of yours will be stretching after something else, still more obtuse and without end."

She was right. Even as he was consumed with the idea of the telegraph, his mind began to return to an old desire. That spring, Mabel, now seventeen, would have a soiree at which she would begin meeting young men. By then Aleck had begun to realize that his feelings for Mabel had shifted. He struggled against it, but more and more he saw her less as a student and more as a woman. The Hubbards asked Aleck to come to the soiree, but just as the pianist.

He came, and he played. For Mabel, he played a waltz, and watched as her young neighbor guided her onto the dance floor. He watched as the young man's hand settled on her waist, that waist Aleck had spent hours training to expand for more air, becoming almost flush with her narrow hips and then releasing its exhale from the delicate dip under her ribs.

Mabel wore a silk peach-colored gown. She couldn't hear the music he played, but she knew the steps. She looked into her neighbor's eyes. She danced.

Aleck was just a poor teacher, just a humble pianist, a family friend, tinkerer, dreamer. Never had he felt it more acutely.

Aleck, exhausted and nearly sapped of hope, temporarily canceled all his private lessons except for George Sanders. Without his lessons with the deaf, though, Aleck ran out of money. Gardiner and Thomas Sanders paid for the invention—they paid for the parts and they paid for Thomas Watson—but they did not pay Aleck. In the spring, Aleck borrowed $9.60 from Thomas Watson, whose pay was $13.25 a week. In the workshop, he had stopped jumping up to perform the war dance he had learned in Canada as he studied the Mohawk language, and which he danced in moments of triumph. Now there seemed to be no triumph. Aleck no longer shouted, "Watson, we are on the verge of a great discovery!," which had always sustained them before. Now they worked in quiet urgency.

On June 1, he had one more lesson with Mabel before he turned his attention wholly to the telegraph. In their hours together, Aleck taught her about the organs of speech.

He communicated to her, at least in part, through one of his many marbled notebooks. Hers was yellow green streaked with orange and had her name in cursive on the cover: *Mabel G. Hubbard*. In it, Aleck explained that the primary issue with her speech had to do with her throat and diaphragm.

He began with the diaphragm: if she could think of the lungs as a sponge for air, then the diaphragm is the muscle that squeezes them, forcing the air into the throat. He began to sketch out a diagram but stopped to beg Mabel's forgiveness—"I do not profess to be able to draw," he said.

He began to outline the lungs, which by his hand looked more like side-by-side small intestines, and the windpipe, the throat. In an adjacent sketch, he outlined, with a more practiced hand, the face in profile: nose, lips, tongue, throat.

On his drawing, as always, the throat was split into three separate pipes, and Mabel asked about these. He explained that the two smaller pipes are the nostrils, the larger one the throat, with a valve that controlled where the air went.

He began to sketch out the possibilities of voice through the nose or through the mouth—it could go one of three ways: only the mouth, only the nostrils, or both. The quality of voice was different depending on which way it escaped, which tube it flowed through. Right now she didn't have good control over that element, but it was easy to learn—it all hinged on this valve: the soft palate.

"If you run your finger along the roof of the mouth," he wrote, "you will find it hard from A to B," referring to the points he had outlined on a new drawing that focused on the interior of the mouth. That whole area, he explained, was the hard palate. Then there was C, the soft palate; "it can move up and down," he wrote, and indicated the movement with dotted lines in a new diagram.

Then he taught her about the uvula, the small strip of flesh that hung down from the palate. He opened his mouth wide and let her peer into it, to notice that by some fluke of nature, he had two uvulas—or more accurately, a bifid uvula, a split uvula.

Then he had her watch his mouth again, as he moved his soft palate up and down. He placed her hand lightly along the side of his nose, to feel the way the sound went through his nose if the palate was hanging down, directing the air through the nostrils.

Aleck made visible and tactile and intellectual everything that hearing people absorbed only as sound. For the most part, he tried to make everything clear. But there were things, too, that he tried to hide.

This work was intensely personal, work of bodies and vulnerability and trust. But when it came to Mabel, Aleck increasingly felt like this was the wrong position for him. In her touch he felt something more; in her trust his heart warmed. He didn't want her to go, didn't want her lessons to end, but he was also realizing that he didn't want to be, forever, her teacher.

The next day, Aleck was back at the workshop. He and Thomas were trying again to send messages across through the latest iteration of their device, but since the receivers were in heavy use and had to be at the exact same pitch as the transmitters to capture the messages, they needed constant tuning. Aleck struggled with it so much that they were now trying only to send three messages at a time to the different receivers. It was Aleck's job, with his musical training, to hold the receiver up against his ear to hear the whine of the current, and adjust the receiver to correspond to that pitch.

In theory, when Aleck allowed a current to flow at one frequency, that current should almost instantaneously vibrate components in the receiver that was tuned to the same pitch, creating a sound. The vibration was converted into a fluctuating electrical current at Aleck's end, sent along the wire, then converted back into sound-producing vibration across the apartment, at Thomas's receiver by the window overlooking Hanover Street.

But now Thomas's receiver was stuck and couldn't convert the electrical impulse. Thomas was exhausted, had no patience left, and plucked at the instrument again and again, cursing it in what he later called "vivid shop language."

At the far window of the next room over, Aleck hunched over his set of instruments, waiting for Thomas to fix his receiver so that he could try transmitting the sound again.

For the past month, even as they continued their work on the multiple telegraph, he'd been thinking more and more about his idea for a telephone. But he didn't think it was worth moving forward on, not right then. The main problem with his telephone idea, he thought, was how distant it was from the principles of the telegraph.

He believed that the power of voice couldn't induce a strong enough current to carry over the wire. Sound breaks down when it travels; it encounters resistance even in the air, degrading its strength over distance. Over a wire it encountered resistance, too. Aleck knew that the voice was strong enough to create meaningful vibrations that could be translated into different forms—it was what he saw with the phonautograph and the manometric capsule—but he didn't think those vibrations could be strong enough to translate into a current that could overcome the resistance of the wire. He thought that in order for sound to transmit, it had to be electrical at the outset, like the dots and dashes of a telegram.

Now Aleck was waiting on Thomas, waiting for the instrument to be fixed so that he could send another electrical impulse and check Thomas's receiver. But then, as he waited, he noticed his own receiver begin to vibrate. It made no sense—there had been no electrical current deliberately sent through the wire, and since the communication was between two receivers, the battery was out of the equation. But as he listened, he realized he could hear the sound of Thomas in the other room, plucking the metal of his receiver.

This was different from anything either of them had ever heard from the telegraph before—not a dot, not a dash, but a faint echo of the "unmistakable timbre" of the other receiver. It wasn't sent by the transmitter but traveled receiver to receiver, with no power but whatever was residual in the wire.

At first he just observed; it was hard to believe that it was real. But slowly he had to believe it. "At every pluck I could hear a musical tone of similar pitch to that produced by the instrument in Mr. Watson's hands, and could even recognize the peculiar quality or *timbre* of the pluck." It meant that a sound—a sound not so distant from the sound of the human voice—indeed had the power needed to travel along the wires.

Even though it was straying from the multiple telegraph that Gardiner had asked for, they repeated the experiment again and again through the day, trying to figure out how it worked and how to make it work again. There was something special about this, Aleck felt, something worth derailing their work for. This sound of plucked metal wasn't as complex as voice, but the fact that it was complex at all gave him hope. He suspected that sound—and by extension, voice—had more power than he'd anticipated.

In order for something like voice to be transmitted, they would need to change the way sound was encoded in the current. To do this, the current would need to remain flowing through the wire instead of being turned on and off. To this flowing current, they would apply a more flexible resistance on the transmitting end, something like a membrane, which could capture the vibrations of the voice and translate it into the undulations of the current.

With a working theory on how to solve the problem of transmission, Aleck was free to think about the secondary problem with the telephone: how voice would be heard on the receiving end. They would need to create a receiver that would respond to more than one frequency. For a simple telegraph, the output of the receiver was of little importance: it could be a light, a beep, or lines drawn on paper. But a machine that carried voice would need a receiver that would respond to a current that didn't simply stop and start but undulated in response to the vibrations of voice. His existing receiver had done this because the sound being sent on Thomas's receiver was tuned to the same frequency as the device that received it, but to respond to voice, it would have to be able to respond to a variety of frequencies—it, too, would need a membrane to capture these vibrations.

The process was infinitely more complex than turning a current into a beep, but it was less distant from an invention that received currents in a number of different frequencies, like the multiple telegraph. Already he was designing a machine that was responsive to variation.

On June 2, Aleck began to sketch it out, his telephone. He hoped Gardiner would be interested, but he also knew that he'd have to be pretty convincing; he'd have to have something real. He sent the drawings with Thomas to the shop, and Thomas began to build.

The next day, June 3, the telephone was ready for testing. They ran its wire from the fifth floor of the shop to Thomas's workbench on the third floor. But even when Aleck shouted with all his might, Thomas could hear only the general timbre of his voice, and a few words. Aleck could not hear Thomas's voice at all.

"A meager result," Thomas wrote. "A bitter disappointment."

But Aleck was not discouraged. He wrote home, "I am like a man in a fog who is sure of his latitude and longitude. I know that I am close to the land for which I am bound and when the fog lifts I shall see it right before me."

Aleck had hoped to be in Canada for the summer, but now there was no end in sight. The multiple telegraph, which was still not working well enough to be commercially viable, was on hold while he pursued this new invention, the telephone, which no one seemed to believe in. And all the while, he received letters from his mother, who was waiting for word that her son would soon come home, where the flowers were late to bloom that year, where the soil was dry from lack of rain. The river adjacent to their property was so small it looked to be drying up, but Eliza held out hope for the promise of a good fruit crop, if only the skies would open. She sent Aleck clippings on rain and motors and evolution and stars. Aleck thought only of vibration, wires, voice.

While Aleck quickly shifted his focus and enthusiasm from the telegraph to the telephone, it wasn't as easy for Thomas. It seemed to him that the telephone only got worse, never better, as their experiments continued. There were problems with the membrane that was key to the capture of the vibrations of voice, and struggles with how to make the encoded current strong enough. By mid-June, Gardiner looked it over but decided it wasn't worth pursuing. He wanted Aleck to focus, to do one thing at a time, and the one thing he wanted him to do was the multiple telegraph. But as soon as Aleck moved toward focus, it was snatched from him by news he couldn't have imagined: Mabel was leaving.

She wasn't going far, just to Nantucket to live with her cousins, but Aleck didn't know how long she might be gone. Several months, at least. He thought it might be permanent. And in the summer's rush of failures—his benefactor's faith waning, his assistant all but given up, his mother calling him home—Aleck's attentions shifted almost entirely. He brimmed not with work but with desire, thinking not of wires and receivers, but of the image of his former student. He found his focus, and his focus was Mabel.

She would be leaving in mere weeks.

Aleck wrote to Gertrude. "I am in deep trouble," he said, before admitting everything: that his feelings for Mabel went beyond friendship, that "I have learned to love her very sincerely," that "if devotion on my part can make her life any the happier, I am ready and willing to give my whole heart to her." He promised that whatever Gertrude wanted him to do, he would follow her wishes—"however hard it may be for me to do so."

Aleck made his way to the Hubbards' home on June 25, the day after he sent her the letter. Gertrude had invited him to join the family for Harvard Class Day, the commencement ceremony that the Hubbards attended along with the city's other luminaries. But first Aleck was guided into Gertrude's sitting room.

Gertrude was clear on what she thought: she judged Mabel to be too young, at seventeen, to know of his feelings. He was ten years older than she, and was her former teacher besides. Though she didn't say it, there was also a considerable power difference here: Mabel was a young deaf woman, still growing into herself; Aleck was a grown hearing man. Gertrude asked him to keep his love a secret for one year.

Thinking of Mabel, Aleck agreed. He did not consider whether this would be possible for him. "My Pride told me I could do it," wrote Aleck. "In whatever circumstances I have hitherto been placed I have always been able to control my emotions . . . and so I went openly into *temptation*."

As the commencement speeches were read, Aleck held his secret close and heavy against him. He made small talk with Mabel and felt he was playing his role as family friend well, or well enough—until Mabel asked him if he ever wanted to be engaged. He didn't know what to say, didn't know *how* to say anything that wasn't a lie. Suddenly, it occurred to him that his secret was wrong, that keeping it was a dishonorable thing to do to Mabel.

He deflected her question, answering calmly, cloaking his feelings. And then he swore never to be alone with Mabel again.

Two days later, June 27, Aleck was alone with Mabel again. It hadn't started that way. It was four days before Mabel would leave, and that evening after dinner, four of them—Mabel, her sister Berta, her cousin Lila, and Aleck—went for a nighttime walk in the garden.

By then, Aleck was beginning to realize that his love for Mabel was more overpowering than he knew what to do with. He worried he wouldn't be able to conceal it. And then Berta and Lila ran ahead of them and out of sight.

There they were, in the garden, Mabel's hand on his arm, and there was moonlight enough for her to read his lips. Though he considered calling out to the others, he did nothing. *Mabel, Mabel,* each heartbeat her name. It seemed wrong not to tell her. The whole garden felt like a held breath.

And then the release: footfalls and rustling and mocking laughter. And Berta and Lila were back with flowers.

They began plucking the petals: *Loves me? Loves me not.* Aleck's flower landed on love and Mabel asked him whom he held in his mind as he plucked the petals. Aleck could say nothing, but Berta and Lila said that they knew, collapsed into giggles, and ran off into the darkness.

Again, just them. He thought of what it would take to tell her, there in the garden, what it would take to wrestle the integrity back into the moment, but then she, too, ran off into the darkness, ran back toward home.

Back on the veranda, Aleck sat at Mabel's feet. Around him, Mabel's family sat with their easy smiles and clean, well-sewn clothes. His own clothes were worn and out of style—everything about him, from his stray hair curls to the way he used words like *pecuniary*, was out of style. He didn't have an easy smile and his velvety black eyes, always, were a thousand miles away. Mabel, slender and long, her eyes bright. Aleck felt how little there was within him that could be attractive to a woman like her. But he thought, too, that he could change, that maybe while she was in Nantucket he could become a man she could love. And then, in front of everyone, her whole proper, upper-crust, Bostonian family, he asked what she was looking for in a husband.

Chapter 7

I feel so misty and befogged. . . . Help me please.

—Mabel Hubbard

A few weeks later, on July 1, 1875, Aleck transmitted the varying pitch of a human voice over a three-hundred-foot wire. The sounds were there, encoded electrically and decoded with his receiver—it was just too weak to be comprehensible. The sounds on the receiving end needed to be stronger; he somehow needed to get more power into the current. "I feel that I am on the threshold of a great discovery," he wrote.

The same day, Mabel left for Nantucket.

By then, Aleck had confessed his love for Mabel to his own parents: "She is beautiful—accomplished—belongs to one of the best families in the States—and has one of the most affectionate disposition[s] that it has been my lot to come across."

Eliza was unenthusiastic. Three days after Mabel left, Eliza's letter arrived: "You are of course the best judge of the sort of person calculated to make you happy, but if she is a congenital deaf mute, I should have great fears for your children. We pray God to direct you my dear boy . . ."

Mabel wasn't congenitally deaf, but the suggestion that it would make a difference was infuriating to Aleck. He stopped writing to his mother.

Aside from the fact that she was in Nantucket, Mabel passed her days as though her life were generally unchanged. Nantucket was an island in recovery from the depression that followed the collapse of the whaling industry, and she explored it for what it was: the newest tourist destination. She went shopping; she went swimming. She spent days with her cousins, confiding dreams, secrets. Her cousin Mary was full of rumors that were as good as truth. One day, in late July, Mary asked Mabel if she would marry Aleck if he asked. Mary, who had never liked Aleck, said she was sure he wanted to ask, and Mabel knew enough to trust Mary's inklings.

Mabel wrote to her mother in a breathless bewilderment, demanding to know if Aleck had spoken to her about marriage. She said she didn't like Aleck as anything more than a teacher: "However clever and smart Mr. Bell may be . . . I never could love him or even like him thoroughly." But she was also moved by his affection: "I am very very much impressed by the honor done me. It is so very strange to think any one would consider me fit to be his wife."

Like her parents, she didn't feel she was old enough yet: "I felt and feel so much of a child still so very young I had no idea of marrying or being sought in marriage yet a while." And perhaps more worrisome, she seemed to not entirely trust Aleck: "I don't feel at all as if I had won a man's love even if Mr. Bell does ask me, I shall not feel as if he did it through love. Any anything but that." But she doesn't say quite what she means. Instead, the letter goes on: at one moment, alarm; the next, consideration.

"Help me please," Mabel wrote, and then she folded, sealed, and sent her letter.

By the time Gertrude read parts of Mabel's letter to Aleck, it had already been days since he'd slept. The next day, he said, he would go to Mabel; Gertrude talked him out of it. But it didn't last, and after just one more sleepless night, he'd changed his mind. He wrote to the Hubbards and said he'd go to her no matter what. "The letter which was read to me yesterday was not the production of a girl," he wrote, "but of a true noble-hearted woman—and she should be treated as such. I shall show my respect for her by going to Nantucket whether she will see me there or not."

On Saturday, August 7, Aleck left all his work behind and boarded the eleven fifteen train to Woods Hole, Massachusetts. At Woods Hole, he caught a steamer to Martha's Vineyard, where he waited all afternoon and evening for the next steamer to Nantucket. He finally arrived late at night and booked a room at Ocean House. For the ninth night in a row, he did not sleep. He believed himself "unfitted for anything."

The next morning, Sunday, he learned that Siasconset, the town where Mabel was staying, was eight miles away from where he was. One man, Captain Baxter, offered rides between the two towns. But he didn't work on Sundays.

Aleck lay down with a headache. He rested until he rose to the sound of rain—the heaviest rainfall he had ever seen. All along the road, windows were filled with faces looking out as the road became a stream, forty feet wide, they'd later say, and three and a half inches deep. Aleck watched the rain, watched as the ground he knew became invisible. Even the air inside breathed it in, the stick and the salt, the smell of wet green things and mud. Aleck looked through the window at a world transformed. He watched the water streaming down the glass, and he thought of Mabel, really thought of Mabel. He began to picture it, him going out to her cousin's home, him knocking at the door, someone going to get Mabel. He thought of everything he had to say, the time it would take for him to say it, and Mabel's face, her eyes as he said it.

Mabel had no hearing and no memory of hearing other than one sound, soft and low, that she called "the singing of the frogs." Unlike Aleck, she had never learned sign language. She understood, instead, through lip-reading, which was hard enough under the best of circumstances. He knew, once he thought about it, that when he began to confess his love for her, her eyes would fill with tears. With her vision distorted, lip-reading would be even harder. He had so much he wanted to tell her, but how much of it could he get out before her tears? And yet, how could he say any less?

Aleck made one last course correction. As the rain beat down, he began to write. A letter was something she could read and revisit; she could feel whatever she needed to feel, could cry, could look away, and she would not lose information—and a letter could slip past barriers the way Aleck, always out of place, could not. And so he wrote and wrote, for thirty-three pages he wrote his love to her, his everything.

He recounted that last night they were together, there in the garden. And then his lapse of self-control on the patio, when he asked what she was looking for in a husband. He remembered the way her family, astonished, held their collective breath, and the way Mabel, unknowing, passed it off with a laugh.

Aleck was grateful for that, for her inadvertent rescue of his transgression, but he was mortified that he'd asked the question in the first place. Still, he wanted an answer. At the end of his letter, he revisited the question; he asked Mabel what it was that she disliked about him, so that he might spend the next year correcting it. He said that if it was too much trouble to tell him directly, that she could send the information through someone else, or, he supposed, not say anything at all. He ended with, "I wish to amend my life for you."

Again, that night, he did not sleep.

By Monday, the rain had stopped, though the storm's gray still lay over the sky; everything above and below was dreary and flat. Aleck got himself a driver and off they went, across the island, along horrible roads, all mud and puddle. The bluish stain of huckleberries dotted the horizon; no trees. They went to Siasconset's local hotel, Ocean View, above the old train station.

Aleck had begun this trip fueled by anxiety, but as the journey went on he thought more of Mabel. He wanted to come clean with his secret, but if she were still as distressed as she sounded in her letter to her mother, he didn't want to exacerbate that. Now that he had written down everything he felt, he could just deliver the letter and give her the time to consider its contents on her own. So from the hotel, Aleck sent a note not to Mabel directly but to Cousin Mary, to say that he was there. He hoped to see Mabel, of course, but he said that she shouldn't feel obliged to see him. He would stay at the hotel until evening, and if he was not sent for by then he would leave. He sent the message, and soon someone came to find him, to bring him to where Mabel was staying.

There, Mary greeted him but told him that Mabel was too distressed to meet with him. It was fine, Aleck said. He gave Mary the letter for Mabel. He asked Mary to tell Mabel not to worry on his account, that he would not bother her anymore until the year was over.

And then he left.

He hailed a ride. He asked the carriage driver to go as fast as he could. From the port, he took a steamer over to Martha's Vineyard, almost missed the one fifteen steamer to Woods Hole, but didn't, and arrived in Boston at seven forty-five at night. He went, immediately, to Cambridge, to the Hubbards', that home that sat on five and a half acres on one of the finest streets in Cambridge. Gardiner and Gertrude had opened that home to him not a year earlier with grace and warmth, befriended him, embraced him. But this time he knew so little of what he would find inside those red velvet wallpapered rooms, knew so little of how Mabel's parents might respond, if they might turn him out.

And he would have to wait, too, after he entered, because Gertrude's mother was still awake and loitering about. The Hubbards sat awkwardly until she finally went to bed. Then Aleck told Gardiner and Gertrude all that had happened. And even though they didn't want him to go in the first place, they now approved of how he'd handled it. Gardiner said that Aleck had done well.

Still, no one knew what Mabel thought.

In Salem, Mrs. Sanders was worried that Aleck was suffering from brain fever—he had been acting so strangely before he disappeared.

Aleck didn't go all the way home that night, didn't take the hour-long train ride to Mrs. Sanders's home. She would have to wait until the next day for her fears to be assuaged. Instead, he spent the night at Crawford House, a Boston hotel not far from his office. For the first night in ten, he slept.

He slept until 11:00 a.m. on Tuesday. Then, he waited.

In many ways, he waited alone. He had broken his promise to Mabel's parents, and he still hadn't written back to his own mother, who had sent her last letter to him over a month earlier and was assuming he was simply too busy to write back. When he finally did write, he explained why he was upset, dwelling on Eliza's use of the word *deaf-mute*. Though both Eliza and Mabel were deaf, neither was mute. This was a social distinction as much as a literal one: deaf people who could speak had more privilege in the hearing world.

On Eliza's use of the word *deaf-mute* to describe Mabel, Aleck wrote, "You would be justly hurt and indignant were anyone to allude to you in that way, and certainly so would she."

He did not revisit the question of heredity. The truth was that Mabel had lost her hearing through illness, and so her deafness itself didn't indicate any hereditary likelihood. But she also had deaf cousins, and though she didn't like to associate with them, it was possible she still carried the gene for deafness.

None of this, at the time, was well understood. Charles Darwin's *On the Origin of Species* was published just sixteen years earlier, in 1859, and the word *genetic*, as in something passed on through genes, was just beginning to come into use. The ideas were new, but they were quickly escalating in social importance. By now the initial backlash against Darwin was dying down, and people were thinking in terms of what traits might be passed down through the generations—not just in animals, but in humans.

"I should have great fears for your children," Eliza had written. At the time, Aleck had taken it as the utmost insult against the woman he loved.

However, sixteen years later, Aleck would stand before an audience of deaf people and issue the same warning: "It is the duty of every good man and every good woman to remember that children follow marriage, and I am sure that there is no one among the deaf who desires to have his affliction handed down to his children."

But that was still to come. Now he wasn't thinking of the fact that he and Mabel were both potentially carriers of the gene for deafness. Right now, his work was simpler: to bring deaf people and hearing people closer to each other, and to bring Mabel closer to him.

On Aleck's return to Salem, he wrote to Mabel once more, to ease any stress she might feel over her need to respond: "Now that I have done my best to show you what I am—I am contented—and can wait. I should only be happier were I sure of your forgiveness and esteem.

"I shall not trouble you any more until the original year is out. And then—if you still think of me as you do just now I shall try to be happy in my work.

"If I may not be any nearer or dearer—believe me at all events

"Your sincere friend, A. Graham Bell."

Days later he wrote to her father, not only about the telegraph, but about the telephone, which he was supposed to have moved on from. He *had* moved on from it, in certain ways, but in other ways the breakthroughs of one informed the development of the other. Aleck believed his most recent breakthrough, the use of a device called a magneto, an electrical generator made of stable magnets spinning on an armature, was key to transmitting electrical information over a distance while retaining intensity. Later, he would realize that the magneto wasn't the solution he needed, but for the time being it pushed everything forward. Wrote Aleck, "I can see clearly that [it] will not only permit of the actual copying of spoken utterance, but of the simultaneous transmission of any number of musical notes (hence messages) without confusion."

The magneto could power the undulating current, and the undulating current could be encoded with much more information at any given moment than an intermittent current. With the use of this magneto generator, either voice or multiple telegrams could be encoded, carried over the existing infrastructure of telegraph wires, and arrive at their respective destinations with enough strength to be reencoded back into sound. He wanted to file a caveat, which would legally declare that he was at work on the invention, or even to alter his existing patent application to cover the magneto-electric current as a means by which to achieve the transference of the human voice, but Gardiner, who was finally becoming interested in the idea, advised against this. He didn't want to show their hand just yet.

Meanwhile, Elisha Gray was in Milwaukee, watching two young boys play with a tin-can-and-string device into which they sent each other their voices. He was also starting to think about wire, vibration, and how to make the voice into electricity. Elisha's inventions for multiple telegraphy may have been neck and neck with Aleck's, but he wasn't thinking of voice with the same level of anticipation or complexity. Sound, to him, was still about the telegraph. Voice was only beginning to play into his thoughts.

On August 14, four days after his second letter to her, Mabel wrote Aleck a distant but polite letter: "Thank you very much for the honourable and generous way in which you have treated me. Indeed you have both my respect and esteem. I shall be glad to see you in Cambridge and become better acquainted with you."

But if Mabel was still keeping an emotional distance, the rest of the Hubbards had started to embrace Aleck. Maybe they were thinking strategically. While under normal circumstances he wouldn't have made the social cut for potential suitors, the Hubbards also must have understood that Mabel's deafness might bar her from being able to marry within her class. Aleck wasn't alienated by her deafness—he was accustomed to the sound of a deaf accent and the small accommodations that would pepper his interactions with deaf people. Instead of seeming alien, these details were fundamentally connected to his mother.

He had, after all, played a role in the way his mother was able to function in public. Because the world around them made little to no accommodations for Eliza's deafness, part of this labor often fell on him. And so being around the details of deafness was familiar for Aleck, was even associated with tenderness and love, though also complicated by the power differential his ability to assist implied.

Now this shifted to how he imagined he would interact with Mabel, if he could be her husband. He imagined that his love would protect her, that it would encompass her like a shield. That so long as he was with her, she could forget her deafness.

It's not clear how Mabel thought of Aleck, what she thought he might bring to her life or the ways he might hold her back, but it's easy to imagine that her parents may have come to imagine that Aleck might bring some kind of social protection to Mabel's life. Maybe they thought no one else would understand Mabel the way he did, would love the differences that her deafness created instead of recoiling from them.

Now Gertrude kept him abreast of when Mabel would be coming home, and to any changes of Mabel's plans. In response to finding out that Aleck had always wished for a sister, Mabel's oldest sister, the younger Gertrude, offered herself as a stand-in, and Mabel's aunt asked if he might want an aunt as well.

It was a true evolution for Aleck, who had long felt a distance of class with the Hubbards. Their trust—especially between him and Mabel's mother—emboldened him.

Mabel came home to Cambridge one day early, on a Thursday, after a scare with the ocean. Mabel always preferred swimming to floating, and she had been swimming on Wednesday, August 18, when she was caught by the undertow. It pulled her under the surface and almost didn't let go.

On Friday, Gertrude wrote to Aleck: "She is well but tired." She invited Aleck for Sunday dinner, but Aleck thought he should wait. He still felt beholden to his promise of a year of silence before discussing his love for Mabel—a promise he had broken and then hastily remade. Now he worried about the way his promise had made him act before. If he was to continue to have Mabel in his life, he didn't feel like he could do with imposed boundaries on what they could or could not talk about. "I have learned too bitter a lesson," he wrote to Gertrude, "to be willing again to lower myself in my own estimation—and in the eyes of other people—even for her love . . . I must be free to do whatever I think right and best—quite irrespective of your wishes."

Gertrude finally surrendered to the idea of Aleck pursuing Mabel. "I believe your love to my Mabel to be unselfish and noble," she wrote. "I trust you perfectly. If you can win her love I shall feel happy in my darling's happiness."

On August 21, Aleck saw Mabel for the first time since she read his letter, nearly two weeks earlier. But it wasn't until the next week, on August 26, that they really talked. Mabel met him at the door when he arrived, and they went to the greenhouse alone together. Mabel told him that she did not love him but she did like him. The rest of the memory is theirs alone. In his diary, Aleck writes, "No need to record here our conversation. Shall remember the whole scene." Her openness, her affection, it was all he needed.

With the question of love finally quieted, he went home to Canada to repair his relationship with his parents. It had been two months since his mother had sent her first letter about Mabel's deafness. In the meantime, they had begun communicating again, but there was still healing to do. Now he carried with him the thought of Mabel, without closure but with hope. There was only time, only weeks, months—perhaps years—of getting to know each other. She wrote to him before he left, "Perhaps it is best we should not meet awhile now, and that when we do meet we should not speak of love. It is too sacred and delicate a subject to be talked about much."

He approached home, too, with his brothers in mind. He knew that he was all the son his parents would have. Even if they hadn't responded to the idea of Mabel as he had hoped, he knew that they were much more upset about the prospect of losing him. And he could not bear to fight with them, to bring them any sadness. They had sadness enough.

He didn't tell them which day he was returning, but his father had figured it out and was waiting for him at the train station. Aleck wrote: "If he has not killed his fatted calf for me he has done everything else to make me happy—and to show his affection to me." He and Melville now felt closer than they'd ever been before. "It is a new and delightful sensation to feel confident that I can approach nearer to him than the outside of his heart," wrote Aleck.

In Canada, Aleck had a horse for his own private use and spent "the greater part of every day" in the saddle. He admired his father's fruit trees and the trellis over which he had trained grapevines. The neighbor estimated a hundred pounds of fruit on the portico alone. And when Aleck wasn't relishing his time with his family, his mind was on patents. He sent detailed, dated notes on his inventions to Gardiner and other scientists—establishing a record of when he experimented with what, leaving a paper trail in defense of future patents.

From afar, Aleck opened himself up to Mabel in new ways. She wrote to him to learn what he thought of Spiritualism. As for her, "I always thought there must be some truth to Spiritualism or it would not be so widespread and count so many adherents, but have never liked to think of it for it seems such a desecration of our holiest and best feelings that the spirits of the dead should return and perform such juggler's tricks as moving and 'tapping' furniture, talking and writing such nonsense."

Aleck fundamentally agreed with her. "What *does* puzzle me," he writes, "is that any sane man can for a moment refer these phenomena to a supernatural source—merely because he happens to be ignorant of any natural cause that may explain them. Strangest of all is the infatuation that can associate these 'juggler's tricks' with the sacred feeling of the presence of the unseen dead." But he didn't dismiss the idea of spirits entirely: "In our utter ignorance of all that passes after death—it would of course be folly to assert positively that the spirits of those we have loved may not still be in existence." And he remembered, too, Melly, and the nights he stayed awake holding séances to try to reach him. "I suppose that spiritualists would say that I failed because I went to work with an unbelieving heart—but I know that I was willing and anxious to give the matter a fair trial."

He began to write teasing letters to Mabel on the subject of women's rights. It was the era of Susan B. Anthony, Sojourner Truth, and Victoria Woodhull, and earlier that year the Supreme Court had ruled that while white women were, in fact, citizens, they did not have the right to vote. Aleck caught wind that Mabel had become interested in the topic: "I never suspected that you were one of these people who think women have rights," he wrote, with alarming coolness—a coolness that he would keep up long enough to distress Mabel. "Do you actually suppose their wishes are to be considered with the same respect as those of men? That their opinions are entitled to the same weight?"

He could feel the absurdity of these statements—and he leaned into it to get a rise out of Mabel—but he also seemed to believe them, on some level. He began to detail the work of women scientists: Mary Somerville, who was "guilty of the most unladylike conduct in daring to write works on the Connection of the Physical Sciences"; Caroline Herschel, who "had the audacity to rob scientific men of the discovery of the connection between sun-spots and the Aurora Borealis!"; as for Susan Dimock, a famous surgeon and developer of the first graded nursing school in the United States—"Words are too weak to portray the enormity of her offences against society. . . . It is true [her work] was to benefit the sick and dying—But are there not men enough in the world to do the work?"

Soon he reveals his hand, "I suppose it will not be long before we have a woman wanting to be President of the United States! Well it is not for me to say her 'Nay'—seeing that I am a subject of Queen Victoria—a woman-sovereign—and one of the best the world has seen—so my best wishes go with her. If women want anything they are sure to get it in the long run—so we better give in gracefully at once."

Mabel's response is brief: "I am a little puzzled to know what your views on Woman's Rights really are."

Aleck, who would spend his life surrounded by strong, opinionated women, was still ambivalent: "I am somewhat puzzled in the matter myself!" On some level, he thrilled at the strength of others, but there was still an underlying unease with the idea of relinquishing his own social power. "My best wishes go to those who try to reform the world—and I should like to help them on," he writes, promisingly, before ending, "*even though they are women!*"

By October 1875, Aleck was back in Boston and Gardiner judged the magneto-powered undulating current far enough along to begin to draft a patent. But despite Gardiner's enthusiasm for the telegraph, Aleck now turned his attentions back to Visible Speech and the deaf. When he turned away from his teaching four months earlier, it was a temporary measure in response to a moment of desperation—he never intended to stay away for long. And besides, his promotion of Visible Speech had created local demand for teachers of the deaf who understood and could implement the method, and that demand was not being met.

While the method was looked at skeptically by other deaf educators of the time, it remained true that most deaf children had hearing parents—and hearing parents weren't generally paying that much attention to scholarly articles on methods of deaf education. What they wanted was simple: they wanted to speak to their children, and their children to speak back to them. Ideally, they wanted their children to hear, but in lieu of that, they wanted their children to be as close to hearing as possible. Aleck's method promised these results; it promised voice. The telegraph, the telephone—they were fulfilling, but Aleck's idea was for them to serve his larger goal, to give him time to devote to reforming deaf education.

Aleck had by now established a reputation for making miracles happen for deaf children, and parents within New England began to seek him out. He wouldn't—couldn't—turn away from that need. But he was also just one man with little money and even less time. He began to devise a scheme that could allow him to push forward the work of Visible Speech and also give him the time and money to work on the telegraph and, he hoped, the telephone.

Beginning on Monday, November 14, he would offer teachers basic training in Visible Speech at his office at Boston University, meeting twice a week for three months for a total of twenty-five lessons. For the complete course he would charge $25. At the end of it, Melville Bell would hold examinations, and distribute diplomas—or so Aleck advertised.

At the same time, he would also teach an advanced course in which teachers would gain actual experience correcting speech as Aleck transferred his own private students over to the instructorship of his teachers in training.

But by late October, when Gardiner found out about the lectures Aleck was giving on Visible Speech, he was appalled. He and Aleck had a business partnership for work on a time-sensitive invention. Gardiner was holding up his side of the bargain, underwriting expenses, but it increasingly seemed like Aleck was only focused on the deaf. "I trust this is not true," he wrote, "because it will . . . confirm the tendency of your mind to undertake every new thing that interests you and accomplish nothing of any value to any one." Gardiner considered Aleck's work since he returned from Canada "a great disappointment to me, and a sore trial."

By early November, Aleck began working on the telegraph again, and determined to set aside some time every day for it. But by mid-month, Aleck realized that he would have a large class for his teacher training school. "Professional visitors are beginning to *pour* in," he wrote to his parents. "All efforts devoted to V.S. just now and to *self*." The fall would be devoted to Visible Speech, and then, in the winter, he could turn to telegraphy without having his profession as an educator of the deaf "slide from under my feet."

But tensions had grown between him and his father over Visible Speech. Aleck wanted to feel himself in a partnership with Melville, and called the advancement of Visible Speech his life's work. But Melville already thought his son had too much ownership of the method. "I regret the changes you have already prematurely made," he writes. "They might and should have been deferred for a new edition in which, as our joint work, all changes might have been made on full consideration by us both. One effect of your premature changes is to render a large analysis perfectly dead stock. You must explain what you want to do before I can give my consent."

Melville's complaints, however, were nothing compared to Gardiner's, who could take it no longer. It is not without irony that Aleck was only continuing the work that Gardiner himself had fought for—a deaf educational movement that may never have taken flight without Gardiner's early support. But Gardiner didn't acknowledge the connection between Aleck's work and his own role in the movement, and neither did Aleck point it out.

Instead, Gardiner was locked onto the multiple telegraph. He wanted it done, but Aleck was not working on it. By late November, Gardiner was out of patience. He gave Aleck an ultimatum: he would have to choose between his work with the deaf and telegraphy. If he chose telegraphy, Gardiner would furnish his living expenses. But money wasn't his only condition; Gardiner brought his daughter into the bargain as well. If Aleck wanted to keep courting Mabel, Gardiner demanded that he turn his back entirely on Visible Speech, his tutoring, and his teaching school. To pursue the woman he loved, Aleck would have to give up the work he loved.

PART
II

Chapter 8

I feel like the first mariner in an unknown sea—uncertain which way to go.

—Alexander Graham Bell

Elisha Gray had long been known for his practical inventions, but in July 1874, when Aleck had only just begun work on his multiple telegraph, Gray promised something wondrous. That day Gray—his hair with a neat part down the side, combed clean, his beard well trimmed—led a *Chicago Tribune* reporter to his Chicago workshop. Despite his recent successes, Gray was still a Quaker farm boy at heart. He was modest, trusting, and even-keeled. He introduced the reporter to his device, which would transmit music: a tiny electric organ, in which the sound was created by electrical circuits instead of pipes. This one had only four tunings—meaning it had only four keys that could be played, four pitches available to send messages—but Gray assured the reporter that this could be expanded to as many musical pitches as existed. His receiver, unlike Aleck's, was not tuned to any frequency. He lifted the receiver, connected it to a wire, and gave instructions to an assistant to wait a few moments and then play a tune.

Gray's telegraphic explorations began when he was just ten or eleven, living on his parents' Ohio farm. He made his own insulated wire from brass wire that he wrapped with linen tape, and concocted batteries from old candy jars. Around 1845, at a time when the telegraph was about as old as Gray was—and Aleck was not yet born—Gray built his first model telegraph station.

Gray's father died when he was twelve, and soon Gray was supporting his family, first as a blacksmith, then a carpenter, a boatbuilder, a farmer, but he always devoted more time to building machines than anything else. One of these machines, a self-adjusting telegraph relay, caught the attention of the Western Union Telegraph Company. The invention solved the problem of telegraph currents losing strength over long distances, which rendered them illegible at the receiving end. Gray's relay, which improved the basic relay that was already a critical component in the telegraph, used a battery to compensate for the wire's resistance. This could give a weak current a new infusion of power, allowing it to reach its destination strong enough to be decoded. In 1868, he patented his invention, and around 1870, he helped start a firm, Barton & Gray, to continue work on telegraphic inventions.

By 1874, the year Aleck began work on his invention, Aleck knew that Gray was competition. Gray, however, wasn't threatened by Aleck. Instead, with a reporter looking on, Gray was quiet and contemplative. Carrying his receiver, he guided the reporter to a room at the far end of the workshop and introduced him to the extraordinary—music radiating from a little tin box.

It was a thing of beauty, but it didn't really mean anything. There was no point to sending music, and so his invention was tamped down to a more practical use. Like Aleck, Elisha was designing improvements to a preexisting invention: the telegraph.

That July, a story about Elisha Gray's device ran in the *New York Times* and was soon picked up by the morning editions of the *Boston Daily Advertiser* and *Boston Daily Journal.* In the story, a Western Union official predicted that "in time the operators will transmit the sound of their own voice over the wires." But Aleck didn't see the story. Aleck was preparing to speak at a convention of teachers of the deaf, held just days later, where he referred to the idea of a speech-reading machine for the deaf. One attendee summarized: "If some simple apparatus could be contrived to bring the vibrations of the speaker's voice to the hand of the lipreader, one half of the ambiguities of lip-reading would disappear." Days after his return home, a note in Melville's diary: "Electric speech (?)."

Back in Chicago, Gray was calling his device a telephone, though it was more analogous to the multiple telegraph. It could send music, which was seen as a neat trick, but though it might have had the possibility of being altered to send voice, this wasn't what Gray was exploring, nor what the patent that he received for the invention claimed. But Gray would soon begin to experiment with voice, and his inventions were beginning to creep into the territory that Aleck was trying to stake out for himself.

Now, in November 1875, as Elisha Gray inched closer to patent rights for his new invention of a "speaking telegraph," Aleck confronted his benefactor's ultimatum. His work with the deaf was where he felt most needed, and he wrote to Gardiner that the "stupendous reforms" made possible by Visible Speech would always be "one of the main objects of my life."

As for Mabel, "Should Mabel learn to love me as devotedly and truly as I love her—she will not object to any work in which I may be engaged so long as it is honorable and profitable. . . . If she does not come to love me well enough to accept me *whatever my profession or business may be*—I do not want her at all. I do not want a half-love—nor do I want her to marry my *profession*!"

The Hubbard household split over Gardiner's ultimatum. While Gardiner had believed that he was doing right by his investments and his daughter, Mabel was less enthusiastic. Insulted by her father presuming to interfere with what she saw as her personal choices, Mabel turned to her mother for help navigating the situation. Gertrude encouraged her not to tease it out any longer, to make a decision about Aleck.

On Thanksgiving Day 1875, which was also Mabel's eighteenth birthday, Aleck visited the Hubbards. Mabel told Aleck that she wanted to marry him, leaving him stunned.

His mother had recently written to him: "You remind me of a fable

I have read about Enthusiasm . . . and Common Sense. The one only looked at the stars and distant mountains—the other the path immediately before him. The former stumbled and fell while the other walked safely in certainty and Peace to the desired goal. . . . Praying that you may take a lesson by Common Sense."

And so for once Aleck deferred to common sense, to caution, and reminded Mabel of all the reasons her parents wanted her to wait: her youth, how little she knew of other men. But she assured him that she knew all she needed to know. Something had shifted in her heart, and she now knew she would never find anyone to love as well as she loved him.

It was beyond what he could understand. She had seemed only to have been moving away from him "—so far away." And now he wrote to her, "I am afraid to go to sleep lest I should find it all a dream, so I shall lie awake and think of you."

He wrote to his parents, "This day—Thursday Nov. 25th—has been appointed by the Governor of Massachusetts—as a Day of Thanksgiving for the Commonwealth—and to me it is truly a Day of Thanksgiving—for Mabel has to-day trusted herself to me and promised to become my wife." He signed with the new spelling of his name, Alec, which Mabel preferred, and which he would use the rest of his life.

But as soon as Alec was gone from Mabel's sight, her certainty turned to dread. She confided in Miss True, now her friend, how unsure she was: "I was so frightened at what I had done." She wondered if she really cared for him, and that fear followed her until she saw him again. In his presence, the tension lifted. She loved him more and more.

Melville wrote to Mabel in approval, and with frankness. "Aleck is a good fellow and, I have no doubt, will make an excellent husband. He is hot-headed but warm-hearted—sentimental, dreamy, and self-absorbed, but sincere and unselfish. . . . With love you will have no difficulty in harmonizing."

In the end, even Eliza was relieved by the engagement, and felt it reflected well on her that her son would marry a deaf woman. In the time that had passed, she backed down from her concern about the possibility of her grandchildren being deaf, instead reassuring Alec that both she and Melville had no prejudices against him marrying a deaf woman: "There is no reason why you and a wife who is deaf should not be as happy as Papa and I have always been." And she was thrilled to learn that Mabel hoped Alec would play piano for her every day of their marriage. Eliza recommended that Mabel rest one end of a stick on the instrument's sounding board and clench the other end with her teeth, in order to feel the music.

She also requested that Mabel learn a bit of sign language, the double-handed British alphabet. "She need not fear yielding to the temptation of using it too frequently, for few besides ourselves understand it." Eliza thought it was important, for safety's sake, to know how to sign a little—to communicate clearly, without all the guesswork of lip-reading.

In late December, Alec set off to Canada to spend Christmas with his family, a long jostling train ride through Massachusetts and New York, circling around Lake Ontario and across the border. As he rode, he dreamed of his future. He did not believe in fate, but in design and work, and so he planned as he dreamed, "cutting and slashing away at my future, trying to mould it into something like shape." The pressures outside of his dream world remained—he knew that Gardiner was still unhappy with his progress, he suspected that Gray must daily be improving his own invention, and, all the while, the demand for Visible Speech teachers of the deaf was growing, the need far outweighing the time Alec was able to devote to training. But for now, Alec was thinking of Mabel.

He made his way from Ontario to his parents' home, arriving Christmas morning. With Alec's arrival, fourteen sat down to a dinner of turkey and goose, the table decorated with holly sent from England by a cousin.

Toasts rang through, one after another, to absent friends, Australian relatives, the doctor, the lieutenant, and then, finally, Alec's uncle rose and asked everyone to fill their glasses to the brim, to toast to "the health and happiness of Miss Mabel Hubbard." All stood, raised their glasses. It was the toast of the day, and the family's voices buoyed these words, *To the health and happiness of Miss Mabel Hubbard.*

The next days unfolded like a dream. Freezing rain covered everything in ice so that the sun glinted off icicles dripping from branches and twigs. Even inside, Alec looked at the scene as though he stood outside of it: clear as anything, his father lay on the sofa, pen in one hand and notebook in the other, his papers and books strewn about. But all the while: "Another scene is floating before my eyes. There is a room within a room—and a faraway picture is before me now." He could see himself, and seated across from him, another figure, long and slender, his future wife. She is writing a letter, perhaps to Miss True. It was, in his words, "as though Cambridge were the reality and Canada the dream."

When Alec returned to Boston, a few weeks later, he began to execute his plans for his future: he wanted to push forward on the telephone and seal his work in secrecy, and so he began to plan for a move from Mrs. Sanders's home in Salem to a $16-a-month two-room apartment at 5 Exeter Place in Boston, where he would set up a lab.

And he was ready to keep his promise to Gardiner, too: he would work on telegraphy from nine to two, four days a week. In his new apartment, he arranged his bedroom in the room facing Exeter Place. In the other room, the one facing the backs of the buildings on Hayward, he brought in a table, removed the carpet "so as to feel more free," and made himself a lab where he could work on his inventions without fear of spies, which were common in this era of invention.

Alec was happy, but Eliza worried that his new landlady wouldn't look after his holes and his buttons the way Mrs. Sanders once did. "I quite fear that you may become a rag-a-muffin when there are no eyes but your own to inspect your clothing," she wrote. Alec didn't worry about this, though. Instead, he gushed over the food. For thirty-five cents, he could have soup, fish, roast beef, a cup of tea, apple pie, and as much bread and butter as he wanted.

As much as he wanted to focus on his work, he mostly thought about Mabel. At first it was a distraction. "I don't know what we are to do when we are married!" he wrote. He was weak for her and worried she would distract him all the time. He recalled a scene of him trying to defend himself in an argument while Mabel, ever so coolly, laid a finger on his lips, closed her eyes in order to stop listening, and simply told him to *Stop*.

"And yet I am submissive!" he writes.

Mabel, who was traveling, responded with characteristic self-assurance: "I am glad you have such a proper appreciation of the benefits of living under my apron strings." As much as he was distracted by her presence, Alec also felt that he couldn't work without her: "I am afraid I shall have no peace when you are out of my sight!" It wasn't until he began to picture her face looking over his shoulder as he worked that he was able to find his focus again.

He returned, then, both to work and to his nocturnal habits. Alec sat in Gardiner's library—the same library where Mabel spent her girlhood earnestly reading though the same clothbound books that now surrounded him, lining black walnut bookshelves. He was working on copies of his latest multiple telegraph patent, which covered the improvements they'd made in the past year. As the night closed in, he erased the months of floundering and tinkering to create the appearance of a final, cohesive theory that was embodied by a working object. But in truth, they were still struggling to complete an effective prototype.

In the new apartment, Thomas, who was considerably less distracted than Alec, ran a wire from the bedroom to the room they were using as a lab, and they recommenced their experiments there. They were focused mostly on the multiple telegraph, at Gardiner's insistence, but the telephone was still on Alec's mind. It had been almost six months since he'd started playing with the magneto—a device capable of giving the vibrations of voice enough power that they could be translated directly into an electrical signal. He drafted his patent to cover several primary elements of both the multiple telegraph and the telephone: the use of an undulating current; the use of the magneto, along with electromagnets, to power the current; and that this device had the ability to transmit vocal sounds. The idea of transmitting voice was radical: at the time, scientists didn't believe that human voice could produce enough energy to be intelligible over a distance. But Alec had the audacious idea that this was surmountable.

As they drafted the patent, they continued to fine-tune the invention. Thomas built the pieces in Williams's shop, and walked the ten-minute walk to Exeter Place for testing, and to design improvements into the night. In the early morning, he would sleep for a few hours in Alec's bed before returning to the next room over, the lab, to work through the rest of the morning and into the day.

During the day, the telephone wires sometimes caught the stray currents of telegraph signals, dots and dashes, but at night, the interference lessened. At night, Thomas listened to their prototype of the telephone, catching the sounds of stray electrical currents and wondering where they came from, what caused them. A snap, followed by two or three seconds of a grating sound, and then it would fade. Another sound, like a bird chirp. They may have been explosions on the sun, he mused, or signals from another world.

Thomas had witnessed signals from beyond before, as a young convert to spiritualism, but these were different: simple cracklings, subtle and cool. In a few years, these sounds would be overwhelmed by power grids, but for now Thomas listened, wondered.

On January 25, a family friend met with Alec, Gardiner, and their lawyers to pick up the British patent application. When this friend arrived in England, which could take as long as three weeks, depending on conditions, he would file the patent. Alec and Gardiner promised they would not file the American patent until the friend was heard from, since filing in America could jeopardize potential British patent rights.

A few weeks later, on Friday, February 11, Elisha Gray handed his assistant a rough drawing and several pages of description for a new invention. He asked him to have a professional draw it out for a caveat (a type of prepatent that legally declared the invention was being developed) and had his lawyer work out the language for it. On the afternoon of the next day, Saturday, the caveat was finished—"Instruments for Transmitting and Receiving Vocal Sounds"—but it was too late in the day to get it notarized. It would have to wait until Monday, Valentine's Day. In the meantime, news of the caveat traveled to Gardiner.

Gardiner, whose entire investment was suddenly at risk, wanted to file right away in order to block Gray's patent. Alec's patent was, after all, finished. It had left for England not three weeks earlier, potentially not even enough time for the voyage across the Atlantic, when Gardiner decided to break their promise. Without consulting with Alec, he prepared to submit the application.

Elisha Gray didn't know he was rushing against Gardiner to get to the Washington, DC, patent office, but he beat him there nonetheless. That Monday morning, Elisha submitted his caveat to the front desk, as was procedure. There, it was put in a basket to be processed into the cash blotter—added to the list of patent and caveat applications for which payment had been received—later in the day, along with the mailed-in patents and caveats.

Later the same day, Alec's patent application for "Improvements in Telegraphy" arrived, but unlike Elisha's application, it wasn't placed in a basket to be processed later. Alec's team knew that time was of the essence, and they had it rush-delivered directly to room 118 of the patent office, the office of patent examiner Zenas Wilber. They cut the line, sneaking Alec's patent in before Gray's.

That both were submitted on the same day flummoxed Alec. "Such a coincidence has hardly happened before," he wrote to his parents.

On February 19, Examiner Wilber contacted Alec's and Elisha's lawyers to inform them of the possibility of an interference—a roadblock that occurred when a patent application was in conflict with another patent or caveat. In accordance with the law, Alec's patent would be suspended for ninety days while Gray could continue to develop his invention and file for a patent, after which Wilber would determine if the two patents were in interference. If they were, all would come down to priority of conception—who had record of thinking of the idea first. There were no guarantees about that; Alec might lose all rights. It was a risk, and Alec's lawyers wanted to prevent the whole thing from going to that stage. If they could, they wanted to secure for Alec what would become the most profitable patent ever granted.

Alec's lawyers, Anthony Pollok and Marcellus Bailey, had connections in the patent office. Pollok lived just a block away and was known to do whatever it took to serve his clients. His partner, Bailey, was friends with patent examiner Wilber. Wilber regularly turned to Bailey for money to support his alcoholism, and was in a near-perpetual state of debt to his lawyer-friend.

The only way to prevent a clash between the two claims would be to prove that Alec filed his patent before Elisha filed his caveat. Alec's lawyers would have suspected that Alec's patent reached Wilber's hands first—they had, after all, expedited it to his office. There was no precedent for using time of day to measure priority in this way, but his lawyers made the argument anyway.

Soon enough, and somewhat mysteriously, the commissioner of the patent office contacted Wilber to tell him that, although neither law nor precedence nor a case he handled just three weeks before had never turned to measuring time of day, they would be doing it in this case.

But since the cash blotter had never been used in that way before, no one had been very careful about how patents were recorded in it. The fact that Alec's patent was expedited meant that it was recorded first, even though Elisha's caveat got to the office before Alec's patent. Nevertheless, Wilber referred to the cash blotter and, on February 25, 1876, sent out letters removing the suspension of Alec's patent.

But Wilber immediately sent out another notice of interference—this one claiming that Alec's patent was in conflict with a patent that Gray had filed on January 27 of that year. Alec's lawyers telegraphed Alec to come to Washington as soon as possible. Alec, who was planning to travel to Washington on Tuesday, February 29, adjusted his plans and left the same day, February 25.

As he traveled, he read over a letter from Gardiner: "You are . . . today worth more than all other men put together—for they stumble on their inventions—you work them out scientifically—you will come out ahead yet so do not be discouraged or disheartened."

Finally, when Alec needed it most, Gardiner was all in. Mabel had said yes to marriage, but he still needed to prove he could make a proper living. There was only this last piece: his invention.

Washington, at the time, was a city strikingly incomplete. In the years since the Civil War, funding for development projects had been diverted to more pressing national needs. Now, at the cusp of the centennial celebration of the nation's birth, the Washington Monument had stood only one-third complete for almost five years, with cattle grazing at its base. It was to this half-formed Washington that Alec arrived on February 26, a Saturday (back then still an official workday). He checked in with his lawyers, and then went directly to see Wilber.

Alec knew that he wasn't legally allowed to see Gray's application—that would be a breach of confidentiality—but he did believe that it was within his purview to know the sections that were believed to be in interference with his own. This wasn't true. The whole thing should have been protected by confidentiality, but certain rules were broken.

Wilber showed Alec the sections where Gray had described something similar to the undulating current described in Alec's patent application.

Alec then edited his application to respond to this. He fine-tuned his description of the undulating current, pointing to an old patent of his that first cited it.

This, too, was a major breach of protocol, not to mention a legal gray area. Nonetheless, Alec's lawyers submitted the amendments on February 29. With them, Alec resolved the second interference, enabling his patent to continue through the review process without having to contend with Gray.

Alec wrote to his father, "If I succeed in securing that Patent without interference from the other, *the whole thing is mine*." He was beginning to see that it could be something big, something more than he'd ever before imagined.

On March 3, Alec's twenty-ninth birthday, the patent was granted.

Back at Exeter Place in Boston, Alec worked in greater secrecy than ever. Alec may have had the patent, but he had it prematurely—he wasn't supposed to have been granted the patent without a working telephone, and he hadn't actually completed his invention. Gray hadn't, either, but his was a caveat application, the appropriate classification for this stage of work. Now that Alec had the patent, the pressure was even greater to make it work. He ordered his supplies one per vendor, so that no one could re-create his formulas. He gave his name only when he had to, and reluctantly. He put all his professional duties on hold, suspended his tutoring, his lectures. He quickly ran out of money. Days passed as they had for months: Alec and Thomas confronting problems, theorizing solutions, incrementally adjusting parts, trying whatever they could.

The main problem was one of gaining enough strength to overcome the resistance of the wire. Alec's use of the magneto to translate voice directly to electrical impulse lacked the ability to amplify the signal sufficiently. Now he shifted gears to work with a different kind of transmitter, which was also covered in his patent. This instrument took the vibrations of voice and transferred them to a membrane with a needle dipping in water. The water acted in the same role as the air and magnets once had, offering variable resistance to a connected circuit and enabling the power of the sound to work with resistance in order to create the undulating current. In theory, this allowed for the amplification that the magneto lacked, though its sound was still quite muffled and quiet.

Then, on March 8, one day after the patent issuance was officially published, Alec's notes reveal a new addition to the variable-resistance transmitter: lacing the dish of water with acid, a feature unique to Gray's caveat, possibly the parts that Alec was allowed to read. By adding the acid, the liquid became a better conductor of electricity, strengthening the current.

On March 10, Thomas finished improvements on this new transmitter and carried it over to Alec's home. They set up the parts: filled a cup with diluted sulfuric acid, connected it to a battery and to the wire that ran between the rooms. When all was set up, Thomas went to Alec's room and stood by the bureau, pressing his hand against one ear and the receiver against the other. In fact, he pressed it hard enough that he detuned it—allowing for a wholly new result.

They'd conducted dozens or even hundreds of these tests, speaking random sentences, clusters of words. If there was one constant, it was that whatever they said was always too faint or distorted to pick up on the other end. On this day, Alec tested the device as usual. He said, "Mr. Watson, come here, I want to see you." But on this day, something different happened. This day, Thomas came running in from the other room and told Alec that he'd heard him—not just some current of electricity, some faint undulation of sound—he'd heard Alec. He'd heard his *voice*.

They switched places so that Alec could hear Thomas, who read from a book. Alec could hear him but described the words as muffled and indistinct. "If I had read beforehand the passage given by Mr. Watson I should have recognized every word. As it was I could not make out the sense."

He heard the words *to* and *out* and *further.* He heard "Mr. Bell do you understand what I say? Do—you—un—der—stand—what—I—say."

By detuning the receiver, Thomas had allowed it to be responsive to a greater range of sounds, like those contained within the human voice. Now they went for weeks shouting words and phrases back and forth, "God save the Queen," and "How do you do?" They understood each other only occasionally, only bits and pieces, the stretches of undulating vowels. They were close but not quite there. They adjusted the battery strength, swapped the needle out for a tuning fork, then two tuning forks. They played with different liquids: cod-liver oil, salt water, mercury, liquor. Even with all these adjustments, it was mostly the case that Thomas understood Alec, whose booming voice was trained in articulation and projection.

It wasn't enough, not for Gardiner, who still thought Alec's work needed all his focus. Alec had been offered a professorship at Washington College, and Gardiner begged him not to take it, to focus all his attention on his inventions. "If you could make one good invention in the telegraph," he wrote to Alec in late April, "you would secure an annual income . . . and then you could settle that on your wife and teach Visible Speech and experiment in telegraphy with an easy and undisturbed conscience."

By early May, Alec had given up on the liquid transmitter and returned to his magneto principle. And by then Mabel had pivoted to her father's defense. In perhaps their first real fight, Mabel stood in her parents' home and threatened not to marry Alec if he didn't finish the telegraph.

Alec felt his heart break. He knew that if she meant what she said, then they shouldn't marry at all, that her love wasn't true. He looked at her firm face and began his own threat—he would leave telegraphy altogether—but he couldn't even finish the thought, and instead he let his vacant words hang in the air, mid-sentence. He left.

That night, from home, he began his letter to her. "I am jealous of your love Mabel. It is your pure little heart that I want—all by itself—and not pinned up as a bribe to the telegraph.

"I want to marry you, darling, because I love you—with my whole heart and my whole soul and I wish to feel that you would marry me for the same reason. . . . It is in the dark hours of life that true love shows itself—and surely you would not forsake me when failure and disheartenment come together—and when of all other times I would need your affections and sympathy."

He wouldn't get a response, a chance to work through the difficulty or to confront the larger struggle it suggested. The issue would simply disappear, at least for the moment. On May 22, 1876, Alec bent down to the receiving end of the telephone and heard Thomas speak—not just vowels, but the until-then elusive consonants. He heard Thomas ask, clearer than he'd ever heard before, "Mr. Bell, are you going to the Centennial?"

Chapter 9

> I shall feel far happier + more honored if I can send out a band of competent teachers of the deaf and dumb who will accomplish a good work—than I would be to receive all the telegraphic honours in the world.
>
> —Alexander Graham Bell

The Centennial had opened in Philadelphia on May 10, 1876, twelve days before Thomas asked Alec over the telephone if he was going. It was a massive world's fair, taking up close to three hundred acres of land, with five main buildings and over two hundred smaller structures set up for the millions of people who would travel through. On display were new inventions like sewing machines and typewriters and the seventy-foot-tall, fourteen-hundred-horsepower Corliss steam engine. There were peculiar objects, too: a tomato-based condiment introduced by Henry Heinz, and the statue Arm and Torch, which visitors could climb inside and stand on, looking out over the fair. They were asked for a ten-cent donation to help the artist finish the piece, which would become the Statue of Liberty.

Alec began his plan for the Centennial almost a year earlier, when he hoped to teach his normal class in Visible Speech to educators at the Philadelphia Institution for the Deaf and Dumb for a $500 salary. He had offered to go immediately, in the fall of 1875, to get things started and planned to "have the work well underway before the exhibition opens." In the end, that plan fell through.

Now, close to a year later, Alec had no plans to attend. With all the time he'd devoted to the telephone, he was now behind on his teaching work. He had papers to grade, and besides, he thought Gardiner could handle this particular telephone task on his own.

Since the day that Thomas had asked if he'd be going, the telephone had continued to improve. But despite their breakthrough in transmitting Thomas's voice, Gardiner still wasn't satisfied. Meanwhile, Alec's own father thought he was working too much on the inventions, and putting off "necessary work in the old style." Alec, too, was having a crisis of faith, and wrote to Mabel: "the more I examine my life and character . . . the more am I frightened for your sake—I do not see there the kind of man that should marry at all."

Alec didn't plan to be at the Centennial, but Gardiner was attending regardless and was determined that the telephone prototypes, even in their half-finished state, should go with him. At that point, the telephone design varied depending on the techniques they were trying, but generally they were made of a large tin cone suspended on a wooden frame. At the small end of the cone was the diaphragm and the magnets, and small wires led to the screws that connected the device to the circuit wires. On the single- and double-pole telephones, the cone was suspended horizontally, and for the liquid transmitter it was suspended vertically over the needle dipping into a tiny jar of liquid. In all these designs the mechanics of the device were largely visible, not hidden inside a box as they would be in later commercial designs. These were the styles of telephone that Thomas replicated for the Centennial, constructing three or four pairs with his best workmanship: he used brass and "polish[ed] them like mirrors."

Gardiner didn't just want the devices, though. They were next to useless without the man who knew how to operate them. From Philadelphia, Gardiner telegraphed Alec. He insisted Alec put aside his other work and come. But Alec was committed to his class of teachers. He didn't want to go.

Mabel was on her father's side. She wanted her fiancé to have the recognition she believed he deserved. On June 17, Mabel persuaded Alec to join her for a casual carriage ride—one that ended at the train station, with

Mabel handing him a prepacked suitcase. He still wouldn't get on the train, but then Mabel—"how pale and anxious she was about it"—began to cry. She threatened again not to marry him, and finally turned her face from him so she couldn't read his lips as he protested.

On the train to Philadelphia, Alec was haunted by Mabel's worried face, by her fear that Gray would get all the credit for the telephone.

"I love you far better than my normal class," he wrote to her, "so I shall try to be patient in Philadelphia for your sake." He promised to do whatever Gardiner thought best, and to stay as long as he advised.

He arrived Sunday night, June 18, and by the next day he was ready to leave. He likened the Centennial to a prison. It was fifty cents to get in, and if you left you would have to pay your entrance fee again. It was a good arrangement for the restaurant inside, but less so for the saloons that lay empty just outside the enclosure, their owners in the doorways, one of whom locked eyes with Alec as he—in a crowd of hundreds—streamed past.

The parts of Alec's various devices were still arriving by mail days after Alec arrived, but it was of little matter since the location of Alec's exhibit was in the educational section, and was mostly populated by the ephemera of Visible Speech.

As he waited for the packages to arrive, Alec wandered through the buildings thinking not about industry or history or medicine or all the inventions on display at the Centennial. He was thinking about his examination papers. He tried to push them out of his mind, quoting Shakespeare to his own thoughts: " 'Avaunt and quit my sight!'—'Let the earth cover thee—Thy bones are marrowless, thy blood is cold . . .' " But his thoughts refused to listen. They broke in, over and over: students, examination papers, diplomas . . .

And so he called Mabel's face to mind, and his worries began to dissipate. When he thought of her, the Centennial felt less like a prison;

instead, he could see the whole world, the accomplishments of its people and nations, a vast collection of wonder.

On Wednesday, June 21, three days after he arrived, he visited Western Electric's exhibit in one of the main aisles of Machinery Hall. There, he recognized Sir William Thomson, a scientist whose work on the transatlantic telegraph had earned him his knighthood; later, he would become known as Lord Kelvin, for whom the Kelvin scale is named. Alec noted that he had "a head and a manner that betoken a man of genius."

Thomson and his wife were examining Elisha Gray's inventions, of which there were many. And as Elisha Gray stood by, quiet and dignified, Alec introduced himself to Thomson, handing him his card. Thomson, in a broad Scottish accent, asked Alec to wait as he examined the Western Electric instruments—if there were time, he would look at Alec's inventions afterward.

But then he became taken by an apparatus—stood over it with such focused attention that he forgot all about Alec, forgot all about his wife, too, until she laid her gloved hand on his shoulder to get his attention. He gave her a loving and pleading look, and she took a seat to let him finish. Twice more she placed a hand on him to remind him of the time, but it wasn't until she'd been there half an hour that he would submit to being torn away.

It was then that he looked back to Alec, startled. Apologizing for keeping him waiting so long, he said he was coming back on Sunday to take a closer look at Gray's telegraph and telephone inventions. Sunday was the day set aside for the judges to examine any exhibit that required quiet. Thompson wanted to know if he might look at Alec's invention then.

Alec hated the idea. Everything that brought his own invention into comparison with Gray's was something he sought to avoid. But on some level he also knew that Mabel was right: it was an opportunity he couldn't pass up. He agreed.

In Cambridge, Mabel was ecstatic to hear about the upcoming audience for Alec's invention, but she still wasn't entirely convinced that Alec would follow through. "I can hardly breathe freely yet," she wrote. And she was filled with missing him. Missing him at night when it was time to shut up the house, missing him in the morning when the room he would sometimes sleep in was left open, the bed perfectly made. But more than missing him, she knew they'd feel lighter when he returned, that the tension the telephone brought to their relationship would fade.

She was attending to her own life, practicing her skills with drawing, a passion which Eliza had rekindled in her. Eliza sent Mabel her own sketches, and Mabel worked on still lifes, writing to Eliza, "I wish my own sketches did not look so rough and coarse beside your delicate and graceful picture." And she was attending to her domestic duties: sewing a white sofa cover to replace a red one and pulling out upholstery tacks, "those horriedest little wrenches," from the wood. Still, she said, she would unearth them all her life long if only Alec could find recognition for his invention.

When her excitement overruled her worry, she wanted to do what her husband-to-be would always do in moments of great excitement: dance the Mohawk war dance. She had never seen the dance, but she could feel it inside, some mysterious beat, some deep desire to shed the weight of her Victorian clothes and manners, some longing to be there with the man she loved, to be present in his anxiety and his joy.

In Philadelphia, Alec surveyed seven smashed glass cells, several flattened telephone cones, and a broken organ. His inventions had barely survived transport, and now he had to ready them to be examined by some of the most eminent scientists of his time.

He worked all day Saturday repairing what he could and getting the devices in order, but he was still nervous about Gray's superior knowledge of battery power, about the fact that Gray had help and support, whereas Alec felt he had none. The only thing he knew he had for sure was *theory*. He had the ability to explain the telephone. And he was a teacher; he could make others understand. Even if the device itself would pale against the immaculate workmanship of Gray's, Alec felt secure in the world of ideas.

The next day, Sunday, June 25, fifty people gathered to see Elisha Gray's beautifully made inventions—including his telephone. Sir William Thomson and Dom Pedro, the emperor of Brazil, both of them judges, stood prominently in the center of the group—Thomson with his eyes always appearing only half open; Dom Pedro, his eyebrows permanently quizzical.

Gray began, but he faltered when trying to explain the science; it was not Gray but Professor Barker of Pennsylvania State University who explained the invention to the crowd. Gray tried, but failed, to send two telegraph messages simultaneously. The crowd was impressed when he was able to transmit musical notes.

They wore full suits, though it was a hot day during a hot summer, and Gray's presentation took longer than expected. Alec's exhibit was the last one, easy to skip. At the end of Gray's presentation, the crowd readied for home when Dom Pedro looked up and spotted an anxious-looking, black-haired man, a man he recognized from a visit earlier that year to the Boston School for the Deaf.

"Mr. Bell," he called, "how are the deaf-mutes of Boston?"

Alec was a man utterly unknown—with this one strange exception of a South American head of state who knew him only for his work with the deaf. Now the eyes of the celebrity judges and their distinguished guests were upon him. Alec went over to Dom Pedro, said the deaf were well, and mentioned that his exhibit—a machine to carry voices—was the next one.

"Well I should like to hear it!" declared Dom Pedro.

Dom Pedro took Alec by the arm, and Alec led the way. The rest of the judges had little choice but to follow—Dom Pedro was an emperor, and it would be rude to end the day without him. So the whole crowd began to walk, and walk.

By the time Alec had acquiesced to being a part of the Centennial, there was no space left in the electrical exhibit; instead, he was exhibiting his telephones alongside Visible Speech within the Massachusetts educational section, in a side room in one of the galleries. "I do not feel comfortable about it at all," Alec had written to Mabel. "So far as I am concerned I do not care one bit who gets the glory of the telegraphic inventions so long as the world gets the benefit of them. . . . If I find that I shall have to smuggle my instruments in—I shall come right back and relinquish the whole thing." But he had stayed, and now suffered the humiliation of leading the judges—and Elisha Gray—to the educational exhibit to witness an electrical invention.

Still, Alec was a man who could stuff down his nervousness, who could appear in control even when his perception of his own worth was teetering quickly between believing himself extraordinary and believing himself a failure. And so he walked with assuredness through the sweltering heat, the emperor of Brazil on his arm.

In the education exhibit, Alec had set up a small table and some chairs, and the group gathered around a number of Alec's devices—the multiple telegraph, the Centennial Single-Pole Magneto Telephone, the Centennial Double-Pole Magneto Telephone, the Centennial Liquid Transmitter, and the Centennial Iron-Box Receiver. Alec's transmitters sat one hundred yards away, at the northeast corner of the building. The night before, he had strung wires along railings to connect the transmitters, and his newly repaired organ, to the receivers.

After explaining his inventions, he demonstrated his multiple telegraph, allowing Sir William and Dom Pedro to message each other. Then he told the gathered crowd about the telephone—warning them that it was an "invention in embryo."

After describing the device, Alec left. He went to the far side of the building, into a room where his transmitter was, a full one hundred yards away. There, he spoke into the instrument: *Do you understand what I say?*

He hoped for two things: first, that sound would be audible; second, that it would sound like voice. But back then, the telephone was one-directional. He would not hear the reaction to anything he said. He spoke his words into an abyss. He waited. He spoke again.

In front of the whole crowd, listening into the receiver, Thomson started to repeat what he heard. *Do . . . you . . . understand . . . what I say!*

Then he jumped up, demanding to see Alec, and set off across the building to find him.

Alec, who was not listening but speaking, talked continuously as the different judges bowed to the device. They heard, "To be or not to be, that is the question," and a constant stream of recognizable phrases. Dom Pedro listened for a moment before exclaiming, "I have heard! I have heard!" Then he, too, jumped to his feet and followed Sir William Thomson.

Alec was continuing his incessant ramble when he heard several heavy, quick footsteps approaching. He looked up, and there was Dom Pedro, approaching at what Alec would later describe as "a very un-emperor-like gait." Behind him was Thomson, and several other judges, all of whom wanted to see what was happening at Alec's end.

Back at the receiving end, Gray pressed his ear to the receiver and heard "a very faint, ghostly, ringing sort of sound," and then, finally, the words "Aye, there's the rub."

Gray thought the instrument looked similar to his own invention, but he didn't worry too much about it. At the time, he was still focused on what seemed like the more practical course: improvements on the telegraph.

It was true—Alec had proven that the telephone could work, that it could be a wonder—but no one thought it was a particularly useful invention. Not even Alec did.

The telephone was still too associated with the telegraph, and as such, it was an inferior invention. The telegraph, with Morse code, was fast and efficient—there was rarely a need to repeat and certainly no reason to have to shout. An operator simply typed the code, and on the other end, another operator transcribed it. It was humble, perhaps, but it communicated clearly and quickly.

People had only imagined the telephone operating along the same lines—not least of all because it had to, quite literally, operate on telegraph lines. They imagined one operator yelling the words to another—and then yelling them again, since the words so often needed repeating. The telephone, on these terms, was a feat of technology, but it served no useful purpose.

Elisha Gray, a few months after the Centennial exhibit, conceded that "as a scientific toy, [Bell's telephone] is beautiful." But he, too, thought of it as operating as a replacement for Morse code: "We can already do more with a wire in a given time than by talking, so . . . its commercial value will be limited so far at least as it relates to the telegraph service."

No one had yet considered that the telephone would render unnecessary central offices of Morse code experts. No one had yet considered connecting people directly, and how that would transform everything: privacy, marriages, work, news. At the time, witnesses could barely wrap their heads around the fact that a voice was emerging from a box. The imagination couldn't fathom all the changes that would be brought on by the power of the voice.

After his demonstration, Alec returned quickly to Boston in order to administer his class's final exams—a measure of how they, too, could summon up voice. But only if he could train them quickly enough to meet demand. He got to work on grading.

At the Centennial, the judges continued to experiment with the telephone, confirming its parts, ensuring it was not mere trickery. Once it was proven that the experiments were successful, Gray approached one of the judges about it, insisting that they were being deceived, that the sound was merely traveling through metallic contact in the wires.

"It's nothing more," he said, "than the old lover's telegraph"—nothing more than tin cans and a string.

Months later, Alec would be awarded a Centennial medal for his telephone, as would Gray for his harmonic telegraph.

But long before that, back in Boston, Alec was continuing to prove that his device was much more than just a lover's telegraph. In the first two weeks of July, Alec began to test his telephone on the telegraph wires of the Atlantic and Pacific Telegraph Company, transmitting music to New York. It was just music, which was easier to transmit because it could be sent note by note—the device didn't have to replicate the undulations of voice—but it was an accomplishment nonetheless. His small lab at Exeter Place slowly became a destination. According to Thomas, the list of visitors during that summer of 1876 "would read like the roster of the American Association for the Advancement of Science." He remembered especially the visit of his old mentor, the inventor Moses Farmer, whose eyes filled with tears as he told Thomas that he didn't sleep for a week when he read of Alec's invention, furious with himself for not discovering it first.

By the fall, Gardiner had convinced Thomas to come work for Alec full-time, and Thomas signed a contract for 10 percent interest in all of Alec's patents as well as the same wages he had been earning at Williams's shop. He moved into a newly vacated room next to Alec's, a room that Thomas suspected had been vacated to flee the noises he and Alec made all night long as they worked. No matter—he and Alec were pleased to further seal the invention in secrecy. They had publicity, and the attention of the public and scientists alike, but they needed to improve the sound quality over a distance. Now they systematized their experiments: adjusting one thing at a time, testing, adjusting that one thing again, testing, testing "magnets with long cores and short cores, hard cores and soft cores, solid cores and wore cores, fat cores and thin cores. . . ."

But even then, Alec's attentions scattered. He was working on planning day schools for the deaf, and opening up teacher training schools in Chicago, St. Louis, and Marquette, Michigan. He was also more in love with Mabel than ever before, always happy to leave the workshop to see her. For his part, Thomas hadn't been in love since he was ten and had forgotten "what an upsetting malady it could be." He described Alec as "quite incapacitated for work."

While Alec quietly succumbed to love, ten million people from around the world circulated through the Centennial, and news of the telephone began to spread globally. Alec was working to perfect his invention, to make it as dazzling as some were now saying it was, but wrote to Mabel, "The telephone is mixed up in a most curious way in my thoughts with you." He believed that the telephone would release him to his happiness, would liberate him to spend time on the things he loved the most. First: Mabel. And second: "Of one thing I become more sure every day—that my interest in the Deaf is to be a life-long thing with me. I see so much to be done—and so few to do it—so few qualified to do it. I shall never leave this work . . . whatever success I may meet with in life—pecuniary or otherwise—your husband will always be known as a 'teacher of deaf-mutes.'"

But despite the rosy ideas Alec had about his impact on the field of deaf education, the truth was that the telephone had caused him to become more and more distant from that work. And in his absence, the Visible Speech alphabet that was central to his teaching was falling out of favor. Teachers found it hard to learn and harder to teach. For students to get the most of the method, it required years of practicing and perfecting individual sounds before allowing speech for communication's sake. In the meantime, students communicated through writing, often drawing out the shapes of letters and words in the air. Even at the Clarke Institution, the teachers were moving on. They admired the quality of speech that came out of the method, but worried about the delay in actually speaking to communicate thoughts. Around them, students continued throughout their educations to draw their ideas in the air instead of speaking them.

Not even Alec thought writing in the air was a good idea. He cared about the perfection of speech, but he wasn't foolish about it. At the Clarke Institution, he'd once taken a fourteen-year-old girl, deaf since birth and with no prior speech training, and taught her to sing a small tune. The teachers gathered around and watched expectantly as the child performed her song. Afterward, Alec turned to the teachers and said, "Now we know she can do it, but it is not worth while, there are too many other things more practical than this to be done for our children."

But Alec believed deaf children should learn to speak, and the early failures of Visible Speech weren't enough to turn away his hopes. While some might see it as evidence of a too-complex system, Alec simply thought he needed more, better trained teachers. Yet the task couldn't fall to him, not then—he was too focused on the telephone.

By October 9, Alec and Thomas had their first distanced two-way telephone conversation, over a line borrowed from the Walworth Manufacturing Company. He had experimented with long-distance telephones that summer, in Canada, but they were only one-way conversations. Now they would try for this major pivot, on which everything depended.

It was a short distance, about two and a half miles, but it worked, and they each kept a record of the whole conversation so that they could compare later. Each wrote down what they themselves said, and then what they believed the other had said. They were, Thomas remembers, "infatuated with joy," and spoke until midnight. Then Thomas wrapped his end of the telephone up in a newspaper, and made his way back to Exeter Place. When he got there, Alec was already gone, left to deliver the conversation to the papers. When he returned, they celebrated, and the next morning their landlord came knocking, threatening to evict them if they didn't keep it down in the middle of the night.

The *Boston Daily Advertiser* printed the results on October 19.

But still, it was only a distance of two and a half miles. The longer the distance became, the weaker the electrical signal became, and the harder it was to understand it at the other end. The telegraph, too, had run into this problem, and at first operators had engaged in a sort of relay to send messages along—decoding them at one station and then reentering them until they reached their final destination. Eventually, Elisha Gray invented mechanical "repeaters," which did the same job automatically. But the telephone didn't work like the telegraph—the patterns of human voice were infinitely more complex than patterns of dots and dashes—and so the telephone couldn't rely on the telegraph's repeaters. The telephone would have to bolster the signal from the source, the transmitter, and that strength of current would have to sustain over the whole distance. Two and a half miles may have been astonishing, but for the telephone to be in common use, Alec and Thomas would need to get it to go much farther.

Thomas was dutiful about waking Alec up at a reasonable hour in the morning, and making sure he went to bed at night. They were close but not close enough, and they were out of ideas. They could make the telephone work over a distance, but not distinctly enough to be of any practical use. If the distance grew beyond their two and a half miles, so would the problems, which became increasingly discouraging. "[The telephone] had got stuck in the mud," remembered Thomas, "and nothing we could do would pull it out." Thomas worked with Alec through the day, and when Alec was off doing something else, Thomas worked on his own: his own methods, his own motivations. He, too, wanted the financial stability of the telephone. He wanted to buy a house for his mother. But he needed a new idea. He needed something outside of himself to split the problem open.

Remembering his childhood séances, Thomas began browsing the newspapers for a medium so that he might consult the spirits on the telephone. He didn't tell Alec. He made an appointment with a woman whose advertisement was the most convincing, but in the end was so disappointed that he never again relied on spirits for telephone help.

Instead, he went to the library. Alec was away and Thomas's mood was sour. At the library, though, he found a description of the magnet used in the Hughes printing telegraph—a "quick-acting" magnet. The phrase stood out. Quick-acting seemed like it might be what they needed. He left the library and was back at the lab as soon as he could be, rummaging through a scrap box to find pieces to the magnet he would have to construct, and then running to Williams's shop for more. It was a magnet that, when he tried it, worked better than any magnet they'd tried so far—the current it produced started strong and stayed strong. This was the piece that made other improvements sing.

With permission from the Cambridge Observatory, he and Alec ran a line from the observatory's Boston terminal to Alec's home at Exeter Place, and continued their experiments over a distance: Thomas stationed at Exeter Place and Alec at the observatory, about four and a half miles away and conveniently close to the Hubbards'. Alec and Thomas identified glitches, tracked their sources, and fixed them. By mid-November, astronomer William A. Rogers at the observatory could tap on his instrument whenever he felt talkative, and if Alec were around and willing on the other end, the two could talk for hours.

They kept extending their distances. By December 3, Alec was stationed in Boston, and Thomas in North Conway, New Hampshire, 143 miles away. Their communication was imperfect, interrupted by storms of static, but Thomas could eventually hear Alec on the other end, singing "The Last Rose of Summer." The *Boston Post* reported: "Professor Bell doubts not that he will ultimately be able to chat pleasantly with friends in Europe while sitting comfortably in his Boston home." Alec quickly filed and was granted patents to cover their advancements.

Now the telephone could carry voices along outdoor wires, across great distances. And more, it now allowed for direct conversation. It had a purpose beyond being a scientific toy, beyond even the telegraph. It needed no codes, no translations. Just two people, a great gulf between them, and a telephone to connect them.

Publicity was picking up, and the telephone was becoming real. In January 1877, Alec began to imagine his invention as a household ornament: "Perhaps one of these fine days you will see Telephones—as a matter of course—in every drawing room!" Even Gardiner, who had fought against the telephone for years, was now ready to embrace it. "It far exceeds anything else you have done," he wrote to Alec, "and has all the qualities necessary for success, i.e., cheapness in first cost, perfect simplicity and almost entire freedom from need of repairs." Not only was he persuaded that it was valuable, he now thought it was the most valuable thing Alec could focus on: "let nothing divert you from this work."

For once, Alec listened to his benefactor. He returned to his usual sleep schedule, working through the nights again. Mabel encouraged him, painting an owl for her nocturnal love. When it arrived at his lab, Alec declared it "my portrait." In the lab, the owl watched as Alec and Thomas worked deep into the night, not going to bed until morning, not waking until noon.

On February 12, the lecture hall of the Essex Institute in Salem filled with people eager to hear the telephone—they filled the lecture room, filled the standing room, the aisles and the stairways, and a crowd had been turned away at the door. The lecture was running late, though, and now the audience banged and stomped and called out for it to begin.

Finally, Alec appeared, walked to center stage, and placed a wooden box on the lecture stand. Rectangular and unassuming, with a circular opening that faced away from Alec, this was the latest design of the telephone. The audience hushed.

By now, Alec had left behind any of his insecurities; he stood tall and dignified, with his black hair and muttonchops framing his slender face and firm black eyes.

He began by explaining his research, as always using demonstration as he spoke. His lecture itself wasn't written out—instead, he simply explained the principles as well he could. Explaining the intermittent current, Alec first ushered from the telephone a sound like a horn, the sound of an intermittent current coming through that receiver. The hall filled with applause. The intermittent current was what was needed to make the telegraph work, and he showed this by demonstrating the telegraphic alphabet. Then he took it a step further, showing how different individual notes could travel along an intermittent current. This was the theory behind his multiple telegraph—that different messages could be sent as different notes or pitches. But for now, he used the theory to play music. He transmitted "Auld Lang Syne" and "Yankee Doodle" on the organ. This was how the old telegraph worked. Now he wanted to show them something new.

Before he started to summon voice from the machine, he asked that everyone in the crowd keep perfectly quiet. Even Mabel, who was in attendance, could feel the silence that followed; she didn't dare to breathe. He then explained the principle of the telephone. This device could, as he showed, do the simple work of a telegraph, but it could do more; it could transmit vocal sounds. He explained this, and then he spoke of his great debt to Mr. Watson.

From the box came Thomas's voice: " 'Hoy! 'Hoy!" And then Thomas began, his voice carrying thirty-five feet across the crowd: "Ladies and Gentleman: it gives me great pleasure to be able to address you this evening, although I am in Boston and you in Salem." The crowd exploded with noise.

For a moment, Mabel didn't know what was happening. There were no lips for her to read, no visual indication of sound from the box. She thought no voice had come, that the lecture hall full of people was laughing at her fiancé. But then she saw their faces were full of naked astonishment; they were cheering, laughing, applauding.

They hushed again as Alec placed his mouth to the receiver. Before he began he felt an "electric thrill" move through the crowd; "they recognized for the first time what was meant by the telephone."

Alec asked Thomas if he heard the applause.

"I wasn't listening," he said. "Try again." And the audience roared.

As they talked, Thomas's voice could be heard throughout the hall—though only those closest could discern what he was saying. Several prominent men in the audience spoke to him.

No one wanted to leave after the lecture, and turning down the gas lights didn't help. Many were still around when Alec dictated a story by a reporter who'd been in attendance to Thomas and a stenographer on the other end, to run in a newspaper that had been founded just five years earlier, the *Boston Globe*. The next day's headline read: "SENT BY TELEPHONE. *The First Newspaper Despatch [*sic*] Sent by a Human Voice Over the Wires*." The story was reprinted all over the world.

In an editorial published the same day, the *Globe* wrote, "Whether the telephone is destined to take the place of the telegraph, or simply to aid it, is a problem for the future. . . . The experiments made in the presence of the Essex Institute last night in Salem caused students of science to stare with amazement at the wonderful effects produced by a simple instrument, manipulated in the simplest manner."

In general, people were shy when they encountered the telephone. They didn't know what to say. It was almost as though they didn't know what could be said, what should be said, how to honor this device with the things they uttered into it.

But more than that, it was a strange thing to speak to a box that resembled an old-fashioned coffee grinder, let alone to expect it to respond. This was a time before recorded music, before radio. Before this, people wrote, or they spoke in person. The idea of speaking to something inanimate, expecting a response, was absurd.

As the story of the Salem lecture spread through regional papers, and then national publications, this started to change. People began to think beyond the absurdity, began to think of the potential for a whole new means of communicating. By March and April, news of the telephone was being covered in London and Paris. Alec was becoming a man of international renown. He was thirty.

Though Alec didn't charge for the first lecture he gave at the Essex Institute in Salem, it was so well received that he decided to charge admission for a second lecture, two weeks later. The $5 per head he charged, which netted him $149, was the first money he ever earned from the telephone. He used $85 of it to order a custom-made miniature: a silver replica telephone for Mabel.

Thomas saw the gift as a waste: "That infatuated young man squandered [his earnings] on a pretty little silver model of the box telephone which he gave to his girl." Cousin Mary weighed in, too, suggesting that instead of a telephone he might have spent the money on a wedding ring. But Mabel kept that silver telephone her whole life. Forty-five years later, she called it "perhaps the most historically interesting thing I have." After the Salem lectures came a lecture in Providence, Rhode Island, and after that the requests filled up his schedule for the spring and summer of 1877. He returned to Boston, then went to New York, and then traveled around the cities of New England. Thomas came along, too, his crucial role to speak into the telephone so the audience could hear. He was uniquely qualified for this, having spent the past two years shouting into the machine. He had developed, in his words, "a vocal power approximating that of a steam organ in a circus parade." But there were times when even Watson's bellow of spoken language was hard for a large group to recognize, and so he also did something that he said could be described, "by courtesy," as singing. The words of a song were recognizable not only by their individual sounds, but by the way the tune suggested those words in the memory of the listener.

The duo's demonstrations weren't foreign to Alec, who had grown up giving Visible Speech presentations, but Gertrude didn't like them. To her they were simply *entertainment*. "I heartily disapprove of Alec's giving telephonic concerts," she wrote. "I think it undignified, unscientific, mercenary—claptrap, humbug, Barnum, etc." She wanted him to start selling the phones and be done with it—none of these crowds. This was the era during which science and entertainment blurred easily; pre-cinema, one was easily the other, and the church often disapproved of them both.

Gertrude wouldn't have to worry for long, though. Come springtime, on April 4, Alec inaugurated the world's first telephone line—not a borrowed telegraph wire, or an experimental wire, but a wire for the sole purpose of using a telephone in daily life. It connected Charles Williams's shop to his home in Somerville.

"I went to his office this afternoon," Alec wrote, "and found him *talking to his wife by telephone*. He seemed as delighted as could be. The articulation was simply *perfect*, and they had no difficulty in understanding one another. The first Telephone line has now been erected *and the Telephone is in practical use!*"

The launch was quiet, domestic. And it would change the world.

Chapter 10

Time and distance overcome.

—Advertisement for the telephone

On July 9, 1877, the Bell Telephone Company was incorporated. Alec, Gardiner, Gardiner's brother Charles Eustis Hubbard, Thomas Sanders, and Thomas Watson made up the board of managers. Sanders operated as secretary, Watson was in charge of manufacturing, and Alec was technically the "electrician," though this role, too, would fall to Watson. There were an estimated two hundred phones in operation, rented out at a rate of $20 a year for a pair connecting a home and another location, or $40 a year for business rentals, plus installation fees. At this point in time, this is how the telephones worked, in pairs connected only by hastily erected wires.

On July 11, two days after incorporation, Alec and Mabel were married. It was a small ceremony in the Hubbards' home, fragrant with Madonna lilies from the garden. As a wedding gift, Alec gave Mabel a cross of eleven pearls, and all but ten shares of his portion (1,507 shares) of the Bell Telephone Company, approximately one-third ownership.

Afterward, they went to Brantford, Ontario, to visit Alec's family, who couldn't make it to the Cambridge ceremony. This was the first time Mabel would meet Eliza, with whom she'd been writing for over a year. In that time, they had continued to correspond about art, Mabel sending Eliza a charcoal sketch, describing her work in oil paints, and admiring Eliza's etchings. When Melville was off on a lecture tour, Mabel wondered in her letters how Eliza was handling it: "I should think you would find the house lonely while he is away." And Eliza had sent gifts that Mabel treasured: first, a beautiful shell, later, Mabel's engagement ring, an heirloom.

Now Alec and Mabel approached through a vine-covered arbor where Eliza waited for them. She hardly let Mabel say a word before she broke a piece of oatcake over her head, in an old Scottish tradition. Mabel thought her "just as nice and kind as she can be, so bright and quick," but quiet, too, so that she reminded Mabel of a little bird.

Alec's family had planned a reception at the house—a party to celebrate both the invention and the marriage. Alec was going to demonstrate the telephone, but as the guests started to arrive, he was still outside, struggling to lay the last of the telephone wire that would connect the home to the local telegraph office. Chief Johnson, of the local Mohawk tribe, arrived to find Alec at work, and quickly got himself a hammer and filled his pockets with staples. Together, they went with their wire along the road, though the woods, orchards, and finally to the telegraph office. Then, as the remaining guests arrived to a spread of ham sandwiches and trifle cake, Alec and Chief Johnson returned to the homestead, joining the thirty guests.

They all hushed as Alec approached his transmitter, lifted the phone, and spoke. After a brief conversation, Alec called over Chief Johnson, "Would you speak in Mohawk for me?" he asked.

And so Chief Johnson took the telephone.

Mabel marveled at the crowd—"such a democratic assemblage I never saw"—as she mingled among farmers, lawyers, drivers, and all the Bells that could gather. It was more diverse and fundamentally different from her Cambridge circles. In terms of class, the Bells were decidedly lower than her family, but they also had a freedom she lacked, a ruggedness and an openness that broadened their circle to people unlike themselves. The next morning she observed eight empty champagne bottles in the hallway as she and Alec readied to attend church on the reservation.

Chapter 10

In early August, a month after incorporation, telephone installation had nearly quadrupled with close to eight hundred telephones in operation, now connecting pairs of houses in Hartford and New York. On the fourth of the month, Alec and Mabel left on the steamship *Anchoria* for their European honeymoon. On board, there were so many Scots that Alec fell into his old accent, adding to Mabel's labor in lip-reading. "I only know I can hardly understand him most of the time," she wrote to Eliza.

They landed in time to make an appearance at the annual meeting of the British Association for the Advancement of Science, where the telephone—and its inventor—became the central attraction. Back in North America, the telephone had reached Chicago, Philadelphia, San Francisco. Melville presided over Canadian rights, setting up lines in Brantford, and Alec made Gardiner the trustee for the rights in Great Britain, Germany, Austria, Belgium, and France.

In October, with thirteen hundred North American telephones in operation, Alec and Mabel traveled to the seaside town of Covesea. It was an old family getaway for the Bells, and now it would be the setting for Alec and Mabel's official honeymoon in a thatched-roof cottage, roses and fuchsia climbing up its sides.

Here they planned to leave technology behind, to live mostly on fish, to be self-sustaining and rugged, though neither of them knew how to cook. They purchased a loaf of bread, a half pound of sugar, and a pound of butter. Alec thought that this would be enough for a week. Mabel was not so sure and was particularly worried about his idea of cooking their own freshly caught fish—neither of them had so much as seen a fish cooked. "Alec appears to think it a very simple operation," she wrote, "but I have an idea that the fish has to be opened and cleaned, and a part of its inside taken out first." But never mind fish, the task of boiling tea and eggs was enough to "break Alec's courage" when it came to food preparation.

When it became clear that they needed help, their landlady brought them fresh milk and then got to boiling eggs, tea, and potatoes for them, eventually arranging to do the rest of their cooking.

Relieved of the duty of gutting fish, Alec dropped lump after lump of sugar into his tea, holding the cup to the light after he did to watch the disruption of little bubbles rising to the surface. Even on vacation, he was an experimenter, and now he was trying to deduce whether it was the sugar itself or the air trapped within the cubes that caused the phenomenon.

They went every day to the rocks, the cliffs, the sandy shore. They stopped bringing food with them, too much trouble, and now they went unhindered. At the shore, the walked along the sand for miles, exploring the caves tucked into the cliffs, keeping an eye on the rising tide. Mabel was falling more and more in love with her husband every day, and in her belly she felt it. It was shifting; it was growing. She was pregnant with their first child.

"O Mamma darling," she wrote to Gertrude a month earlier, "if I only could put my arms tight around you and ask you if it may really, really be true what Alec and I are just beginning to hope may be. I'm most afraid to write it for fear it would be so hard to feel that you will go on hoping after it is all over—the hopes and fears. And I am so afraid that it may be all wrong . . . Mamma, you must not say anything or believe anything until I write again." But day by day, her certainty grew.

Climbing back over the cliffs, Alec set up blankets and shawls, leaving Mabel with the company of a book while he went off "hunting." Mabel wasn't fooled. For all the show of it—he could wrap a stone in paper, throw it in the air, and shoot it on the way down—she was pretty sure he spent more time just watching the birds. She described his pistol as "quite harmless."

They walked alongside farms, the air without a hint of frost, so "soft and mild" that the harvest had not yet been taken in. They passed countryside crisscrossed with wire fences, and Alec bemoaned all that wasted wire, which he thought could be perfect for the telephone. Even now, on his honeymoon, he couldn't entirely give himself over to rest—his mind was always alive, always obsessing. Soon he and Mabel were back out there with his device, he connecting it to wire fencing, she yelling with all her might into the transmitter. The fact that it didn't work didn't dissuade him. It was just that the wire had grown rusty, he said.

He went back to watching birds, telling Mabel all about his hope to build a flying machine. His thoughts darted everywhere. "He goes climbing about the rocks and forming theories on the origin of cliffs and caves," Mabel reported, "which last problem he has solved to his satisfaction. Then he comes home and watches sugar bubbles, starts out the next morning after rabbits . . ."

Their last night there, tired of his too-short bed, Alec slept outside on the beach.

In the morning, Alec wrote a poem in the sand with a mussel shell and danced wildly around the lines as he wrote them. When he got stuck on a rhyme, he stood with his bare feet wide, his arms outstretched, his face puzzled.

Mabel worked on sketching him, marveling at her husband's mind, moving without a moment's notice from torpedoes to telephones, tides to ruins to flying machines. She hadn't found him handsome when they met, nor even though their courtship, but now she did. She saw him in so many new ways now. When they'd first met he seemed so careless, improper, "hardly a gentleman." But then she saw him in Tutelo Heights and in Covesea, and he seemed the height of class compared to the ruggedness of the people he grew up around. Somehow he'd become perfect. And now he was at the center of what was becoming one of the most important inventions in the world, and he loved her. He stood before her on the sand, arms outstretched, feeling the power of the sea behind him, the puzzle of words before him.

Then, turning to the sea, a look of shame crossed his face. It didn't seem proper, he said, that the inventor of the telephone should go wading. Mabel was full of love for him, for the family they were now beginning.

She looked at her husband and shrugged. "Don't be a slave to your own position," she said. And with that, he set out into the sea.

After their honeymoon, Alec and Mabel traveled from Scotland to England, and when she was certain of her pregnancy, they put off their return to the States—originally planned for November 1877—until October 1878, on the advice of their doctor. In the meantime, they rented a seventeen-room house in South Kensington, living far beyond their means. Alec continued to work on the telephone, giving lectures that drew crowds of as many as two thousand people, and coming up with small but vital improvements like pairing the two circuit wires within a common insulation, to decrease interference on the line.

As the pair settled into their married life, Alec's fame grew. "Everybody seeks to do him honor," wrote Mabel. "He has been introduced to all the great people, Lord this and Sir that, and all are anxious and eager to speak with him."

Despite all the effort Mabel had poured into keeping her husband on track to becoming a famous inventor, she now shrank from the attention. Every new person she met would have been a new communication challenge, a sap on energy. She wrote to her mother: "Oh, I wish I were safe at home, I'd give up all hope of ever being Lady Bell and all the glories of being the wife of a successful inventor, everything but just being Alec's wife, to come home to quiet humbleness."

But Alec's star was rising. In January 1878, he got a letter from Queen Victoria's private secretary asking when it would be convenient for Alec to present the telephone to the Queen. He wrote back that the next Thursday or Friday would be good for him, but that he "awaited Her Majesty's commands," a particularly British suppliance that disgusted Mabel.

But that's not to say she wasn't excited—it was, after all, the Queen—and she ordered a Parisian dress for the occasion. She tried not to be too descriptive in her requests for it, asking for something simple and not showy, for receptions and ceremonies, for this new life of hers, as the wife of a famous inventor. She anxiously awaited it as Alec drew up phone lines and arranged for famous singers to be posted on the other ends. He had a telephone receiver made for the Queen, with polished walnut and gold switches.

Mabel's dress arrived soon after she found out that she wasn't actually invited to the event, and she saw it through the veil of her disappointment: its black silk, its excessively long train, its horrible beaded trim, its exorbitant price. "Well," she wrote, "if this isn't a showy dress, I don't know what is."

Meanwhile, on the evening of January 14, Alec waited for the Queen. He had grown out his beard and put on thirty-six pounds since his wedding day, weighing in at 201. Since his pants kept tearing, he'd arranged to have new clothes made before visiting the Queen. Now he stood with a solid grandeur, imposing and dignified, if a little less light on his feet.

He waited in the Council Room of Osborne House, a royal retreat on the Isle of Wight. The room was furnished in white, red, and gold, adorned with three-quarter-length portraits of Queen Victoria and Prince Albert. She arrived at nine thirty in black silk and a widow's cap, the mourning garb that she'd been wearing since her husband's death sixteen years earlier. At first, Alec found her "humpy, stumpy, dumpy," her physical appearance fat and her hands coarse, but he soon found that she was warm and dignified and he quite liked her. He spent about fifteen minutes explaining the invention to her, showing her a telephone that had been sawed in half, so she could see the inner workings. The Queen was entranced by him.

And then, after Alec had secured the phone connection, he began to hand the phone to her, only to find that she was looking away. He, who spent his days around deaf people, lightly touched the Queen's arm to get her attention. But nobody touches the Queen, especially not some scrappy inventor. There must have been a collective sharp intake of breath, a pause of horror, but the Queen just turned and took the phone when he handed it to her.

They listened to a singer, Kate Field, on one line, and then later switched to another, connected to Cowes. As the concert songs began, the Duke of Connaught exclaimed, "It sounds like distant music on the water!" and soon took to the line to talk with people on the other end. They called Southampton and then London, and around midnight, the experiment ended.

Queen Victoria called the invention "most extraordinary," and the newspapers had a ball with Alec's faux pas, drawing a picture of him pulling at the Queen's arm as she beamed.

Alec and Mabel settled into their temporary London home, and on May 8, 1878, Mabel gave birth. The first time Mabel held her daughter, the child looked up at her with "grave wondering questioning wide open eyes." Alec examined her mouth, her eyes, her ears, verifying that all were functioning well. He wanted to name her Darwinia, for Charles Darwin, but he was voted down. Instead, she was named Elsie, a Scottish variation of the name Eliza, for his mother.

Alec took to his newborn daughter with his usual curiosity. "He is at once so fond of it and yet so afraid of the poor little thing," wrote Mabel, "and he hardly knows how to hold it." As it was, Alec awaited the day when she would seem real to him, old enough for him to feel like he could truly love her. "I often take her up in my arms and nod to her and play with her from a feeling that it's my duty to try and like her—but somehow or other if the truth were told I have to imagine what she'll be like when she is two or three years old before I can summon up any *real* feeling of affection for her." He was quietly tortured by this—"at my secret hard-heartedness"—but he was tortured, too, by any strange bubbling in Elsie's mouth, or by any "suspicious noise" her body made.

Outside the walls of their home, the telephone was becoming

ubiquitous. "Wherever you go," wrote Mabel, "on newspaper stands, at news stores, stationers, photographers, toy shops, fancy goods shops, you see the eternal little black box with red face, and the word 'Telephone' in large black letters."

By mid-November, the English press was running even the tiniest telephone stories, and the public flocked to whatever lecture on the subject they could find. People began experimenting with the telephone not only across great distances, but also *down* great distances, into mines and attached to the helmets of divers. They worked to connect police departments with each other; they worked to connect banks. Telephone caricatures graced the pages of *Punch* while physicians tried to see if the device could help them understand more about respiratory diseases.

In America, it was satirized. The *Daily Graphic* in New York City ran an image titled "Terrors of the Telephone—the Orator of the Future," featuring a speaker, sweaty and crazed, speaking into a receiver whose wires stretched to audiences in China, Dublin, Fiji, and dozens of distant lands around the globe. In this image, it's not peace or reconciliation that voice brings, but *too much* connectivity. This one man, his frenetic energy radiating through the wires, is able to reach anyone, anywhere. This new technology was too much to handle, the image suggested; we shouldn't be so connected. Some stillness, some peace, was being lost.

Chapter 11

I do not see in this school the two or three children actually present—but the thirty thousand deaf mutes of Great Britain.

—Alexander Graham Bell

In Great Britain, as his daughter entered the world, as the telephone grew in popularity, as everything in his life felt like it was coming together, Alec finally found himself free enough to return more wholeheartedly to his work teaching the deaf. It was the winter of 1878–1879, and Gertrude wanted him to use this time to resume his speech lessons with Mabel—she feared Mabel's skill was slipping, that it needed to be further refined. Alec refused to do it, insisting that he and Mabel, as husband and wife, were equals, and they couldn't reenter into the unequal power dynamic of teacher and student. But he wasn't turning his back on teaching altogether.

Over the past year he'd been corresponding with a man, Thomas Borthwick, who was starting up an oralist school in Greenock, Scotland. Alec started advising him in November 1877; in May 1878, he traveled to Scotland to give a lecture to support the founding of the school; and then he began to arrange for a teacher to come from America to help with the school's launch. Alec loved the idea of the Greenock school, and thought it could serve as a pilot program, "likely to inaugurate a revolution in the method of teaching deaf mutes in Europe." And he was ready to turn his full attention toward this work again.

As it was, in Scotland, if people wanted their deaf children to have an education at all, they had to send those children to England or Europe, removing them altogether from their families. This removal concerned Alec. When the students returned, he thought, "they were looked upon as strangers. The only friends they had were amongst themselves." This, he believed, was how it would remain all the rest of their lives. "In large cities the adult mutes dwelt in communities, intermarried with one another and tended in that way to perpetuate their defects."

But as he pushed forward in the idea that this separate community was undesirable, he refused to admit the ways in which the deaf community created a space in which deaf people could access their own sense of power and self. Instead, his insistence that deaf people were in no way different than the hearing led him to think that deafness—and the language and community that blossomed within it—offered nothing. He saw deafness as a "defect," and suggested that the world would be better off without it. This school, like all oralist schools, would be built off this assumption.

The Greenock school was slated to open in early September 1878, but by late August, the teacher Alec had arranged was nowhere to be found. And so Alec began to travel back up to Greenock himself, to temporarily act in place of the missing teacher and open the school.

When Alec imagined these children whom he would be teaching, only five and six years old, he grew worried. It had been years since he had been in front of a classroom. He feared that they might not defer to his authority, that they might run to their mothers. He imagined potential tantrums: children screaming and crying on the floor of the old lumber room where the school had been relegated. Red-faced, tear-streaked, they would beg to be let out of the room.

Alec's fears proved unfounded: the children were bright, pleasant, smiling, ready to learn. At the start of the day, they gathered between piles of books, slates, pens, "deceased writing matter," and "bundles of mysterious build," all of which were covered in dust and cobwebs so old they were blackened. It was humble, but it was just the beginning. And as he waited for permission from the trustees to clean, he drew a face on the blackboard.

This is where they began, how he'd always began. There weren't many students throughout the course, just two or three on any given day. But Alec didn't mind. For him, it wasn't only about the students present; there were thirty thousand deaf people in Great Britain, all of whom he believed stood to benefit if he could succeed with these children.

Here in Greenock, Alec worked to educate the children in speech without the use of signs. He began with an easy word, like *pea*. First he wrote it on the board in English, then in Visible Speech. When students didn't know written English, like the youngest, he would use gestures to guide them, like any foreign language teacher would, but there remains a question of what additional signs he may have used. He wasn't opposed to the manual alphabet, and with his own fluency in ASL, it's hard not to wonder if the language slipped in. However he did it, he guided the children to their own pronunciations of the word. As they spoke, he wrote what they said, in his symbols, underneath the word they were meant to say. They began to learn how close or far what they said was to the word they were supposed to say. He made a game of it: Who would be the first to offer the correct pronunciation?

He was back doing the work that he saw as his calling; this was the return he'd sought for years. He wanted to bring his ideas forward, push them out into the world, and uplift, as he did, the whole of the deaf community. All deaf people, he thought, could be empowered to go out into public with self-assured dignity, could speak to anyone they wished.

Alec envisioned not just a school but a revolution, a liberation. Yet he'd been away from the work for years now, and within the world of deaf educators and deaf people themselves, the arguments around oralism had shifted. It wasn't just about the declining use of Visible Speech; by the late 1870s, a decade after Clarke opened, the whole method of oralism was falling out of favor. Its biggest argument—that speech was necessary and doable—had been embraced by the manualists, who had increasingly incorporated speech into their ASL-based educations. Now many of them were calling themselves by a new name—not "manualists" but "combinists," or the group that combined the best opportunities of each method. But *combining* didn't mean *conceding*. They didn't believe that signs needed to be forbidden in order for students to learn speech.

Their argument now was much bigger than Alec realized and had much greater stakes. By now everyone had seen what oralism could do—and what it couldn't. Namely, it couldn't teach the deaf to speak perfectly, and only a small number would even learn to speak intelligibly. And what was more, while speech was being taught, basic education seemed to lapse. Both sides understood that, as valuable as speech was, acquiring it required sacrifice. Now the argument was over how much to sacrifice. It seemed like the students learning by oralist methods were considerably stunted in their emotional and intellectual growth. Was speech worth that?

For the most part, deaf leadership—teachers and principals of deaf schools, writers and publishers of deaf newspapers, and founders of deaf clubs and associations—was excluded from the highest levels of this conversation. The best they could hope for was to have the attention of one powerful hearing man: Edward Miner Gallaudet. Informed in part by his communications with deaf leaders, he brought many of the deaf community's concerns to the forefront of the debate. And he would present Alec with his greatest challenge as he moved back to his own work with the deaf.

Edward was no newcomer. He was following in the footsteps of his father, Thomas Hopkins Gallaudet, who had helped introduce sign language to American deaf education in 1817 after his trip to England and France.

When Edward took over his father's work with the deaf, he did so with the understanding that sign language was fundamental to the education and empowerment of deaf people. His own mother, Sophia, was deaf. As an adult, Sophia couldn't speak—but that didn't get in the way of her life, not as Edward saw it. Edward believed in the equalizing power of sign language.

When he opened the first college for the deaf, the Columbia Institution for the Deaf and Dumb (later the National Deaf-Mute College and today Gallaudet University) in 1864, he envisioned it as a place where deaf people could "be made the social and intellectual equals" of hearing people. At a time when even relatively few hearing people went to college, the college extended an affordable opportunity for talented deaf people to access higher educations—often becoming the first people in their families to do so. The college fundamentally shifted the possibilities for deaf equality.

In 1867, the same year as Gardiner opened the Clarke Institution and Melville published *Visible Speech*, E. M. Gallaudet went on his own trip to study deaf schools abroad. There, as he watched a variety of methods in a variety of European schools, his pro-sign language approach became more complicated.

At an oralist school in France, he found that the sacrifices required for an oralist education were considered worth it. For students, "the power of communicating freely in speech with their fellow-men . . . was so great a boon as to justify a lower standard in the intellectual training of the deaf and dumb." This alone might have shocked Gallaudet, but it was nothing compared to the knowledge that the teachers felt this concession was justified despite the fact that only about half of their students ever mastered speech well enough to use it in practice. Of the students Gallaudet spoke with, some he could understand clearly, others not at all.

Again and again he saw the students struggle with general education. In one school, where the students' speech was very well developed, he found the simplest of math equations being taught in the advanced classes. There seemed an inverse relationship between the advancement of speech and the quality of education.

But it wasn't just that the fundamental educations of these children were sacrificed—something even more insidious was going on. In one German school, a high-achieving child was brought to Gallaudet to show him how she could pronounce certain elementary sounds. She did well with the vowels, but struggled with the consonants, repeatedly getting them wrong until she began to cry. Of this incident, Gallaudet wrote, "the acquisition of artificial speech is oftentimes, under the most favorable circumstances, a painful and embarrassing task to the pupil."

Gallaudet began to see what speech could do for those who had learned it, but he also saw the harm it could do both intellectually and emotionally. He didn't take this lightly. He was an early voice advocating for manualist teachers of the deaf to begin to instruct in articulation, but his recommendation was couched in the idea that the training should be abandoned for any student who didn't seem well-aligned with it, and that it should never stand in the way of general education. Speech was valuable, but not as valuable as the oralists claimed: "The utterance of many pupils was so indistinct and imperfect as to be understood only when most closely attended to, while that of others was, to a stranger's ear, hardly more than gibberish." He didn't judge the speech of these students to have been worth the work that must have been put into it.

Back in 1874, while Alec was beginning to pivot his attention more fully to inventing the multiple telegraph, Gallaudet shifted his focus to defending sign language, which he felt was increasingly under attack. In an 1874 *American Annals of the Deaf and Dumb*, the most prominent journal of deaf education at the time, Gallaudet relayed his impressions of a visit to the Clarke Institution—the school that Gardiner had started and still the flagship school for the method. Of the students Gallaudet observed, he was impressed by some: those who had been taught for six years and were able to speak in such a way that they could be understood by a stranger.

He would come to believe that the ability to speak with this level of fluency was worth whatever work was put into it, and would not have been possible under a fully manual, or sign language–based, method. However, he was careful to express just how few deaf people were able to attain these results. He cited the numbers given to him by a German teacher of the deaf who had taught articulation for fifty years. According to him, in a group of one hundred students learning orally, eighty-five students could converse with the people closest to them; sixty-two could do so easily; and only eleven could converse easily with strangers. This was in line with assessments from other teachers he'd communicated with—only about 10 percent of deaf students would ever be able to speak with easily understood fluency.

Gallaudet came to see articulation training not as a necessity; rather, he saw it as a way of playing to a skill, akin to nurturing a student's ability in poetry or music. It wouldn't make sense to insist that all students become poets, and neither did it make sense to insist that all deaf students speak. Teachers could give lessons to those students who showed a special talent at it—the skill could certainly prove useful—but to insist that all students learn to speak was to risk the rest of their educations.

What Gallaudet and other combinist teachers were arguing was, essentially, that these students needed whatever education they could get as swiftly as possible. The problem with oralism wasn't that it was impossible; the problem was that it took too long and that the results were too meager.

In the nineteenth century, no one was talking about neuroplasticity, about the way the mind of a young child was more malleable than that of an adolescent or adult; however, putting speech at the forefront carried tremendous risk for the development of the brain. From birth to about five years old, a child has the ability to develop a native language, and this language acquisition lays pathways in the brain not only to understand language but also to understand things associated with language: communication, of course, but also abstract thought, problem-solving, empathy, and social interaction. It doesn't matter what the language is; what matters is that the language is fully accessible to the child before their language acquisition window closes.

The problem of oralism wasn't just, as Gallaudet and others believed at the time, a problem of speech training taking classroom time away from general studies—that was, unfortunately, a vast understatement of the damage. Language is a fundamental building block of the human mind; it was almost impossible for the mind of a child without language to process and understand anything else they were supposed to be learning, either intellectually or emotionally. Language had to come first, and language wasn't coming.

English was not a fully accessible language to most deaf children, and the time they spent floundering to learn it was time that their language acquisition window was closing. If they missed the window, they became language deprived. They would never get that neuroplasticity back, and would likely never have the ability to fully learn any language. Today, this is understood by experts as irreversible brain damage.

Gallaudet may not have understood this, but he could see evidence of what was happening—he could see the intellectual and emotional struggle of deaf children educated orally, and he was conscious of what the deaf community was arguing based on personal experiences. The deaf community was largely united for the preservation of sign language, and Gallaudet argued this, too. He said that deaf kids should be exposed to sign language immediately, and taught the foundations of written English through lessons conducted in sign language. Sign language was the most equitable, efficient, and compassionate vehicle for the education of deaf children.

In 1878, as Alec was turning back to the issue, oralists were doubling down. They had begun to argue for the absolute abolishment of sign language in the classroom. Alec didn't practice this himself, but he didn't stand against it, either. Some oralists argued that the learning of more than one language was injurious to the child, and almost all oralists argued that ASL was an inferior language. This wasn't Alec's argument, either, though he did believe that learning sign language stood in the way of social advancement.

While many manualists believed that only a small number of deaf people benefited from oralism, some oralists insisted that close to 99 percent did. The numbers cited fluctuated wildly, depending on measurement and what one considered success, and while it would be easy to find an article to correspond with whatever one already believed, it was also true that the vast majority of the articles published argued that oralism primarily benefited those deaf people who either became deaf postlingually or who retained some of their hearing. Oralism, for them, was rehabilitative speech therapy. It was building on something they already knew.

These were the people Alec was most familiar with—people like Mabel and Eliza, who were deafened postlingually. And he was drawing broad conclusions about all deaf people based on what he saw in this subset. This was happening even in Greenock. There, Alec was teaching ten children, but he wrote that he saw the whole of the deaf population of Great Britain in them. He said this as he observed only what was possible from the students who grew up in the most ideal conditions for speech training, students who already had language.

Alec was used to people telling him that certain things couldn't be done. Now he ignored the possibility of failure, but in doing so, he was also ignoring the harm that might be done by not admitting that he could be wrong. Now that he was the inventor of the telephone, his ideas were received differently by the outside world. Everything he believed was imbued with a sense of believability, even genius.

Teaching again at Greenock, even briefly, relit a spark in Alec. He was ready to reenter this world. He was ready to devote himself fully to the oralist cause.

For the first time since telling Gardiner about his idea for the telephone, he was able to focus fully on his true work. He was happy at last. "Happier," he wrote, "than at any time since the telephone took my mind away from this work."

Mabel had stayed in London, but now Alec begged her to come join him. They could live in Gourock, he said, two miles down the coast from the school, a town he described as "mountain and sea blended together."

She didn't want to join her husband in Gourock. She didn't want her husband in Gourock at all. Later in life, she would admit being jealous of the attention Alec directed toward deaf children, and questioned his attitude toward them, wondering if all deaf children were, on some level, only "cases" to him. And at this point she wasn't so far removed in time from when she was his student.

Mabel wanted nothing more than to forget deafness, to live her life liberated from it. It was the promise of her speech-based education. It allowed her to identify as not-quite-deaf. It allowed her to move closer to "normal." The problem was that proximity to deaf people threw the whole act out of balance: "to have anything to do with other deaf people instantly brought the hardly concealed fact [of my own deafness] into evidence." And it felt to her that her husband kept pulling her back to this world she wanted to leave.

She had resisted his work with the deaf before, but back then the resistance was, on the surface, about him focusing more on the telephone. Now something else was emerging. It seemed she didn't want him working with the deaf at all.

Not that it mattered much to Alec. "You would really laugh to see how happy I am in my work," he wrote to Mabel. "I like teaching little children far better than working at the Telephone. If I could only be relieved of pecuniary anxieties—I should be perfectly happy to devote my life to this work. I do so love little children—and I like nothing better than being among them."

Mabel wrote that he was spending all his time with "a couple of babies."

Alec snapped back: "You are mistaken in supposing that all the pupils are 'babies'. . . . I do not consider myself as working so many hours a day 'for a couple of babies'—but as inaugurating a revolution in the methods of teaching deaf children in this country. . . . I am not going to forsake my little school just when it is struggling for existence—though the telephone should go to ruin—and though my wife and child should return to America and leave me here to work alone."

His response was extreme, and it didn't end there. He threatened to stay in Scotland until Christmas, if that's what it would take for the school to succeed, but then softened into the heart of the matter: "It is a sorrow and great grief to me that you always exhibit so little interest in the work I have at heart."

He was willing to leave her, at least temporarily, for the work. Or that's what he said in one breath; in the next, he said he missed her. These were his two greatest pulls, his greatest loves, and he couldn't stand them in conflict. He couldn't sustain it even for the length of a letter. Their time in Covesea was still so near to him, rambling on the beach, stringing telephones to fences. When Alec thought of Mabel, all he really wanted—deep in his bones, deeper than any argument—was to lay his head on her shoulder, to feel comforted by the presence of his wife. To wrap his arms around her, feel her love, give love to her.

Mabel traveled to him in mid-September. There, as Mabel wrote letters to family, baby Elsie sat on her lap, tearing up pieces of paper. Elsie was turning out to be a beauty, a good traveler, a good sleeper. The mothers of Alec's students took quickly to showering her with gifts: a silver mug, a Scottish dress.

Alec had wanted to stay in Scotland with the Greenock school, but the permanent teacher eventually arrived, and it was time for him to return to his life, to the past he was running from. Alec wanted nothing more to do with the telephone, that was clear to everyone. For now, the Bells would travel back to North America together, a family, in only a few weeks. They planned to go to Canada, to spend some time with Alec's parents, before moving on to Washington, DC.

Alec and Mabel were reunited, in love, building a home and a life. But there were things that were not resolved, fundamental wedges between them: her resistance to the fact of her deafness, everything she had learned to be and to hide, her aloneness in the experience of this struggle; his inescapable devotion to his life's work, a part of his heritage, his reason for being, and his inability to see how this rubbed against his wife's identity.

Chapter 12

They have hacked it to pieces. They have torn it limb from limb—They have plucked out the heart of the invention and have thrown it away . . .

—Alexander Graham Bell

On November 10, 1878, Alec and Mabel arrived on the Quebec shore. There, in the midst of the waiting crowd, Alec made out an unexpected face from Boston. It belonged to someone who was now older, richer, and whose dreamy boyishness had released a self-assured man. Thomas Watson's muttonchops were now a full beard, his scrappiness polished up, the part in his hair clean and strict. But like Gardiner and Sanders, the security of his future was still dependent on the telephone, on Alec. And the sight of Watson meant Alec could no longer run from the thing he'd been trying to escape.

Alec hadn't even wanted to return to this continent—let alone see faces from Boston. Despite the telephone's growing fame, the Bell Telephone Company was struggling. They should have had a monopoly, at least for a while, but instead they had competition. In December 1877, nearly a year earlier, Western Union had organized the American Speaking Telephone Company to bring together the work of three of the other inventors working in the field of telephony: Elisha Gray, whose harmonic telegraphs were marketed under the Harmonic Telegraph Company; Thomas Edison, who had invented an improved transmitter; and Amos Dolbear, whose telephones were distributed under the Gold and Stock Telegraph Company.

The American Speaking Telephone Company brought all these

companies and inventors together and hooked their telephones up to the telegraph lines Western Union had all over the nation. They leased these telephones to the public, in potential violation of the Bell patents and undercutting the Bell Telephone Company's rates.

Not only that, they had started a larger publicity campaign. They claimed that Alec had not invented the telephone, that he stole the invention from Elisha Gray, that the patent office was in Alec's pocket, which was the only reason the telephone rights were granted to him. A piece ran in the *Chicago Tribune* on February 16, 1877, which claimed that Elisha Gray was "the real inventor of the telephone."

Soon, the American Speaking Telephone Company was outselling the Bell telephone. Western Union had existing infrastructure, $3 million in annual profits, and the new Edison carbon transmitter, which allowed voice to travel further and clearer than with Alec's transmitter. Bell Telephone may have had claim to the telephone, but Western Union had claim to the improvement that made it considerably more functional. The companies began talks toward some kind of consolidated company, which would be financially safer than hashing it out in court, but again and again, they couldn't come to an agreement.

Those who had poured their money and time into Bell's business were growing frustrated. Thomas Sanders had invested $110,000 in the Bell Telephone Company, and still hadn't seen any returns. The company was $30,000 in debt, the company's employees hadn't been paid in weeks, and suppliers were starting to pressure the company for money owed to them. In America, public opinion of the Bell Telephone Company was low; stock prices were lower; and the finances of the company were slow to stabilize—they seemed to spend as much as they earned from the telephone. Thomas Sanders's finances had been exhausted by early 1878, around the time Queen Victoria reached out to Alec. If the company didn't find a way to get out from under Western Union's growing power, there would soon be no company at all and—Queen or no Queen—the money lost would never be recovered.

In early September 1878, the Bell Telephone Company sued the man in charge of telephone distribution for Western Union, Peter Dowd, for patent infringement. In Dowd's defense, they expected that Western Union would argue that Alec didn't have rights to the telephone, that he didn't have priority of invention. This would give the Bell Telephone Company the opportunity to establish Alec's rights in court and stop Western Union from infringing on the telephone.

But that wasn't the end of the Western Union problems. While Bell was still in Scotland, Amos Dolbear, an inventor with Western Union, was working to disrupt everything the Bell Telephone Company had built. He believed that he had, in fact, had priority of conception over Alec in the use of permanent magnets and metal diaphragms in the telephone design, which he had theorized in September 1876. With Western Union's help, he applied for a patent to retroactively cover his use of a permanent magnet in the telephone, thus creating a patent interference with one of Alec's patents. Alec would need to make a preliminary statement in response, explaining his claims to the invention and the dates of his work—to essentially block the interference. But despite letter after letter from the Bell Telephone Company, Alec had not yet responded. He had been focused on that deaf school in Greenock.

All Alec wanted from the telephone was that it give back what it had always promised: enough money for him to go on with his real work. This, it still wasn't giving, and so he swore it off. "Good-bye to it all," he wrote, "I lay it bye in my past."

But there were still men whose fates were entwined with his, men who wouldn't let him off the hook. The Bell Telephone Company executives were tiring of Alec's resistance. They needed him to represent his invention: to defend his patents and drum up financial support.

Alec and Mabel had chosen to sail on the *Sardinian* into Quebec instead of Boston precisely so that Alec could avoid the Bell Telephone Company. They planned to travel to Tutelo Heights and visit with Alec's parents, at the house whose walls were covered with Eliza's paintings, whose furnishings were old and solid and familiar. They would eat the fruits of the recent harvest and revisit Alec's dreaming place above the Grand River. And he and Mabel wanted to introduce his parents to their granddaughter, Elsie, with her swarthy skin and shock of black hair and her father's wandering, wondering eyes.

But Thomas Watson had been sent by Gardiner to catch Alec as soon as he was off the boat and promptly deliver him to Boston. Those at the Bell Telephone Company saw it as their last shot. Even though the court case hadn't begun, there was a time limit on Alec's statement to defend his patent against Dolbear's interference. If Alec missed this deadline, it didn't just mean his patent would be vulnerable to Dolbear, it would also make victory in the Dowd case, and thus victory against Western Union, all but unachievable.

On the shore, Thomas found Alec even more resistant than he anticipated. Alec said he was done with the telephone. He was going to be a teacher of the deaf. Thomas, though, explained his own bind and slowly wore Alec down. Still, Alec said he refused to put any more money into telephone work. Would the company pay for his travel to Boston? The telegrammed response came the same day: yes. Alec finally agreed, but only if he could go see his parents first.

"I went with him," wrote Thomas, "because I didn't want to lose him."

Together, they took the train to Tutelo Heights, Alec growing faint and complaining of stomach pains the whole time. As soon as she saw Alec, his mother called the doctor, and then settled down to some time with her first granddaughter. Alec was soon diagnosed with two abscesses. The doctor lanced one, but couldn't treat the second.

Days later, still weak and pale, Alec boarded a train to Boston while Mabel stayed behind. Upon arrival in Boston on November 14, he went first to a hotel, and then to Massachusetts General Hospital. He insisted to Mabel that this arrangement was basically just a hotel room with doctors—they would be applying a poultice to try to bring the abscess to the surface. In truth, he would have surgery the next day. And he still hadn't made his preliminary statement regarding the telephone.

That same night, in Chicago, Elisha Gray approached Highland Hall. The ornate building was lit bright, and orchestral music spilled from its doors and into the neighborhood of Highland Park. It was a reception and banquet put on by his Highland Park neighbors to honor the person they believed rightfully deserved the title "inventor of the telephone."

At the banquet itself, where Gray had requested no alcohol be served, the guests raised glasses of water in their toasts, and then they sat back to listen to several honored guests tell the story of Elisha Gray's invention.

One of the arrangers of the event, S. R. Bingham, began: it had all started a little under five years ago, around 1873, when the neighbors first started whispering among themselves that Gray had found a way to transmit music over wires. By the summer of 1874, he gave a small demonstration in his home, followed in December by a public demonstration at the local church. He steadily developed this device over the course of the next year, until February 14, 1876, when he filed for his caveat and collided with Alexander Graham Bell. While Gray explicitly stated in his caveat that the ultimate goal was to allow vocal conversation over telegraph wire, Bell's patent had nothing to do with speech, they said, and yet through some mystery of patent-office loyalties, he had swiped the invention out from under Gray.

Gray listened on, his face strong and yet reserved; "it is not difficult," wrote one reporter, "to imagine he is engaged in close thought upon some problem of his own."

The speech went on, arguing for the priority of conception that Gray clearly had for the telephone. But patent or none, Bingham said, the public was beginning to recognize Gray. "His fame shall be assured," promised Bingham, "and the children shall read, in the school-books and upon the monuments, the honored name of Elisha Gray." Time would correct the error of the patent office.

The day after Gray's reception, November 16, doctors at Massachusetts General Hospital operated on Alec. It was supposed to be routine, but nearly a week later he was still there. Doctors told him they were going to need to operate again, to open the wound and let the inside heal while the wound stayed open. This time Alec feared both the pain and the ether. This time, he refused surgery until his wife was by his side.

Though Gardiner and Alec still hadn't completely repaired their business relationship, Gardiner was there for his daughter and her family. When Mabel said she wanted to join Alec, Gardiner traveled to Canada to escort her back to Boston.

Now, with Mabel holding his hand, Alec was put under. Despite the pain, he was quickly out of the worst of it. From the hospital bed, while still recovering, he at last dictated to Thomas Watson his preliminary statement defending his rights to the telephone. It was delivered to Washington, DC, just before the final deadline expired.

"The worst is over now," Mabel wrote to Eliza, "he has but to lie quiet and all will be well."

But the quiet only opened up all the worries inside of him. It must have been dawning on him just how precarious his claim to the telephone was. What had happened in the examiner's office two years earlier was wrong—Alec should never have been able to look at Elisha's caveat—but it was clear from Alec's letters at the time that he didn't know that. The people around him surely did, but he wrote to his parents that it was "my right to see the portions of Mr. Gray's specifications which came into conflict with mine." There's no record of when Alec realized he shouldn't have known what he knew about Gray's caveat. But in the hospital, as he was faced with a return to the telephone, he was filled with anxiety.

Alec couldn't get a solid grasp on what the future would hold for him. *What is to come next?* he wondered. *Where are we to live? How are we to live?* Alec went over the options for his life: he still wanted to be done with the telephone but felt indebted to Gardiner; he wanted to teach at the National Deaf-Mute College but feared he wouldn't be able to make the kind of living his wife desired; and someplace deep and unspoken, there must have been that dread that he was about to be found out, that his telephone patent wouldn't stand up to investigation. He had seen the papers, and he knew, with a sinking misery, that while he lay in the hospital unable to even eat, Gray was being honored in Chicago. If the telephone collapsed, everything collapsed—not just for him, but the Hubbards, the Sanders, Thomas Watson, Mabel, Elsie. The anxieties and the unknowns were overwhelming. But now Mabel was here, her body soft and familiar; she put her arms around his neck, and he thought: *All will be okay, somehow, in the end.*

"It is too wintery here!" Alec complained of Boston in the early weeks of 1879. Alec, now recovered, had stayed to assist the Bell Telephone Company's legal case. In Boston, temperatures held steady below zero, with ice and snow everywhere, "even in the telephone office." Those who worked there, now including Alec, were "shivering over the ashes of the Bell Telephone Company." But it wasn't just the shock of the cold that ravaged Alec's composure; he was also watching in horror as his lawyers dissected his patent—"My beautiful patent of 1876!"

"They have hacked it to pieces," he wrote to Mabel. "They have torn it limb from limb—They have plucked out the *heart* of the invention and have thrown it away . . ."

By now, Alec had new lawyers, specialists in patent law. Chauncey Smith was the Bell Telephone Company's regular counsel, and a neighbor of the Hubbards' in Cambridge. With shoulder-length ringlets of gray hair rounding out his face, he had an old-fashioned largeness about him—he was heavyset with just as large a personality, by turns belligerent but also joyful. In certain ways, he wasn't unlike Alec.

The company's other lawyer, James Storrow, wasn't like Alec at all. A graduate of Harvard Law, he was small and quiet with a closely trimmed beard. He studied technology carefully and thoroughly, his mind homing in on the smallest details of things.

Together, they looked at every sentence, every claim, every specification of that nearly three-year-old patent. Their job was to rescue it. Alec's job, though he was loath to do it, was to stay out of the way and follow instructions.

There was no more talk of teaching at the college; instead, in exchange for Alec putting off his own career to be on hand for the legal defense and to further develop the telephone, he was paid $3,000 a year by the Bell Telephone Company and $1,000 each by Gardiner and Thomas Sanders. As the Bell Telephone Company faced bankruptcy, Alec was fully engaged working to prove ownership of his invention and save his reputation.

On the afternoon of January 21, he went to William's shop and began to build the telephone that he described in his patent. He should have had to provide a working prototype to the patent office when he first filed the patent, but he never had, and the requirement was waived. Now he wasn't even sure if the invention he had patented would work.

At the time, the way that courts were considering invention was in flux, driven in part by the American image of an inventor as a lone discoverer, a cowboy, a hero. The inventor was always a single person with a dramatic breakthrough, whose credit was never shared. In truth invention was a long slog toward a quiet finish. Inventors built on the ideas of those who came before, and built, too, on the ideas in the air, the whisperings of what other inventors were up to.

These two notions of what an inventor was were in conflict nowhere more vividly than in patent law, and in the later half of the nineteenth century, the number of patents was exploding: in 1850, fewer than one thousand patents were issued. By 1870, six years before Alec filed his telephone patent, that number was more than twelve thousand. But what exactly these patents meant, legally, was still in question.

If inventions were a slow confluence of many ideas from many people, then what—and whom—would a patent protect? Should patents be small in scope, a continuous patenting of small improvements? Or occasionally did an inventor-as-hero emerge, one who could sweep all the rights for the invention for years to come?

As always, America preferred the hero. But to be a hero, you had to have the type of patent that definitively struck out on new ground—a pioneer patent. It couldn't be a small improvement over prior inventions. There had to be a eureka.

Alec knew his eureka. It was simple, plain, and beautiful all at once: the undulating current. It allowed for something other than dots and dashes to be transmitted. This was what he had described in his patent application back in 1876, and by January 23, they had finally built the exact invention as Alec described. "The instruments *talk perfectly*," wrote Alec with a sigh of relief, "and are only inferior in *volume* to the latest forms of instrument." Three years after he patented it, his discovery had been proven to work exactly as he imagined—the undulating current, *eureka!*, there at its center.

But Smith and Storrow were not just lawyers, they were storytellers, and they knew something radical and important: everything hinged on the spoken voice.

That was the eureka in their eyes, the major leap forward, the thing that qualified his patent as a pioneering patent, with all the wide sweep of rights it would carry. They thought that Alec's undulating current was a small progression, something that would get bogged down in minutiae. But if Smith and Storrow designed their argument around the transmission of spoken voice, if Alec had been the first to lay claim to a device that could *speak*, then he could be the hero they needed.

Alec *had* made this claim. But it was a small claim, his fifth and final claim in the application: that the patent for this invention would secure "the method of, an apparatus for, transmitting vocal or other sounds telegraphically, as herein described, by causing electrical undulations, similar in form to the vibrations of the air accompanying the said vocal or other sounds, substantially as set forth."

Alec hated his lawyers' idea. It stripped out all of his scientific work, all of his true discovery. "They have subjected my invention to hydraulic pressure from legal minds and have squeezed out *into the gutter*—the very life-blood of the idea. All that remains of my poor specification is—a *little dry dust*—which they blow in my face as the essence of the speaking Telephone!"

As much as Alec objected, though, his lawyers stayed the course. Their plan wasn't only about ensuring that Alec's patent was framed as a pioneering patent—it was also about throwing the other side off their game. Smith and Storrow suspected Western Union's case was being built around the idea that Alec wasn't the first to generate and transmit an undulatory current—that they would have spent all the intervening months sharpening this very argument. It would be an easy argument to make—or at least it would be easy to dredge up enough inventors working in electricity to throw considerable doubt on Alec's claim. Already, they had the stories not only of Gray, but also of Thomas Edison, Amos Dolbear, and Johann Philipp Reis, all of them inventors who had done considerable work in electricity and the transmittance of sound and whose work preempted Alec's in a variety of ways. In American patent law, the earliest inventor still trumped whomever got to the patent office first—if they had the documentation to back it up. All Western Union had to do was prove that someone else was there first. This would undermine the primary claims of Alec's patent, centering on undulating current.

In Boston, on January 25, Alec's lead attorney, Chauncey Smith, opened their case, claiming that Dowd was infringing on the company's telephone patent rights. They called their single witness to the stand, Henry B. Renwick, a civil and mechanical engineer and former patent examiner. Renwick immediately admitted that the telephone was built off the unoriginal concept of undulating currents. He argued that the original idea that the patent sought to protect was the transmittance of speech—something no one else had before claimed to do.

Smith asked Renwick a total of eleven questions before ending politely, "Gentleman, that is all."

"I never saw men so taken aback and surprised," wrote Alec of the opposing side. "To their astonishment . . . the invention they have stated to be old is not the invention we claim they have infringed and they don't know what to do." All of Alec's annoyance and resentment toward his lawyers melted into glee. "To their surprise [the defendants] find they have been waiting and preparing for an attack that is not to be made—and a sudden assault has been made in an unexpected quarter—and in the most feebly protected spot. Before they can have time to change their front—and before they can bring any cannon to bear upon us the chances are their fort will be taken—and we will be in possession of the ground we want."

Alec's side may have launched a surprise attack, but there was still a long trial ahead. During February and March, the Bell Telephone Company strengthened from within by merging with the New England Telephone Company. Together, they formed the National Bell Telephone Company. Alec rallied for Gardiner to helm the combined company, but Colonel William H. Forbes took it over, and management of the National Bell Telephone Company slipped almost entirely from the Bell Telephone Company owners. Gardiner remained involved to some extent, staying on the five-person executive committee; Thomas Sanders had already resigned his position as treasurer, but he, too, joined the executive committee; Thomas Watson, young and now rich, began to move on to other ventures—he would try his hand as a farmer before pivoting to building steam engines, founding the Fore River Ship and Engine Building Company in 1884; and Alec finally received one piece of his request to put the telephone behind him—he was left off the executive board.

The same action that Alec once thought would free him—removal from the helm of the company that bore his name—actually meant that he was now fully freed to devote his time to his invention's defense. This was Alec's life now: trial. Work with the deaf was replaced by meetings with lawyers and appearances in court. Alec split his time between Boston, where the Bell Telephone Company was headquartered and where the case was being heard, and Washington, DC, where the Hubbards had moved, and where he and Mabel rented a home at 1509 Rhode Island Avenue, within walking distance to Mabel's parents.

As Alec traveled between the two cities, Mabel sat in his study and wrote to him. She had cleaned the study, then immediately wished she hadn't, that she'd left the clutter of her husband to surround her. "I miss you dreadfully every moment," she wrote from his desk. Even the rare moments when she felt okay would crumble when she thought of just how often and how long he would be gone, "and then my heart and courage go down into my boots."

The separation wore on Alec, too: "Don't let us consent to being separated any more; help me darling to prevent it now. Let us lay it down as a principle of our lives that we shall be together, that we shall share each other's thoughts and lives—and to be to one another all that a husband and wife could be. Letters cannot speak as we can face to face—heart to heart."

In March, Alec was back in Washington with Mabel again, but even then he was preparing to leave for Boston, reading through old letters for anything that could be used in court. His secretary copied anything he found, and Mabel packed and unpacked Alec's trunks. Elsewhere, Thomas Watson prepared models of telephones for the trial. In the coming months, Alec would present his own testimony.

April 4 was the first day of Elisha Gray's cross-examination. Alec felt bad for Elisha—"for I feel sure that he would never of his own accord have allowed himself to be placed in the painful position in which he now is."

On April 7, Alec's counsel brought the court back in time to February 1877—a pivotal month in the history of the telephone. In Alec's world, this was only weeks after his lecture in Salem, his first lecture on his new invention, in which he called Thomas, who then spoke to the whole of the packed lecture hall, garnering the headline "Salem Witchcraft" from one paper.

It was more than just a big month for publicity, though. It was also the month when Elisha Gray came to terms with Alec's contribution in a documented way, providing a key piece of evidence in Alec's case. As Alec had basked in the glory of his Salem lecture, Elisha Gray prepared for a lecture at McCormick Hall in Chicago, where he would display and explain some of his inventions, including his telephone, as well as Alec's telephone, a model of which he could have built within a day's notice if Alec gave him permission.

Days before the lecture, Gray wrote to Alec, telling him that he intended to display the telephone, and to give him credit. "I should explain it as *your* method and not mine," he wrote, even though they had both been successfully working toward the same basic invention.

Alec consented, as long as Gray also explained to the audience the libel of the *Chicago Tribune* article that ran earlier that month, which credited Gray with the invention.

Gray hadn't read the article, but he could guess what it said and didn't want to forfeit the claim that he, too, invented the telephone—even if he didn't get to the patent office first. "So far as I know," wrote Gray, "the 'libels' are mostly on the other side, if assertions of originality, etc., may be so considered. The papers here have been full of articles, of late, copied from Boston papers, claiming the whole development of the telephone for you."

Alec responded apologetically: "For of course," he wrote, "you are not responsible for all the ill-natured remarks that may appear in the newspapers concerning me." He went on to clarify what he did give Gray credit for—advancements in multiple telegraphy—and to explain why he felt the telephone invention justly belonged not to Gray but to himself.

As requested by the court, Gray read the letters aloud: his first to Alec, the reply, his second to Alec, the *Chicago Tribune* article referred to, and Alec's reply to the second letter. None of this was any surprise to Gray—he remembered writing these first two letters and was prepared to respond to them. But there was a third letter from Gray.

In that letter he wrote, "I gave you full credit for the talking feature of the telephone . . . in my lecture in McCormick Hall." He pointed out that this acknowledgment made it into the Associated Press dispatch, which was printed across the country. And he wanted to clear the air, too: "Of course you have had no means of knowing what I had done in the matter of transmitting vocal sounds."

Gray's work toward transmitting vocal sounds was theory, though, and he closed the letter with: "I do not . . . claim even the credit of inventing it, as I do not believe a mere description of an idea that has never been *reduced* to *practice* in the *strict sense* of that phrase should be dignified with the name invention."

This was the quote that could destroy him. *I do not claim even the credit of inventing it.*

In court, the words "fell upon Gray and his counsel like a thunderbolt," wrote Alec. Before Gray even began to respond to questions, he turned to his lawyer and said, "I'll swear to it, and you can swear at it!" He knew the letter was true, that it was what he'd written. He also knew how damning it was.

Legally, this wasn't a smoking gun. What Gray believed in the past about the inventions wasn't necessarily on trial, but for his image and credibility it was a swift punch to the gut.

The case was far from over, but by the summer, Western Union's chief counsel told the company that Alec's patents couldn't be defeated. He contacted Smith and Storrow to begin discussing a compromise.

Alec gave his own testimony in July, a testimony so thorough that it filled almost one hundred pages. At one point, when Alec had been speaking from midmorning to two in the afternoon, one lawyer finally interrupted to ask if Alec might want to pause for lunch. "I don't lunch," said Alec. He continued talking.

The testimony was superb. Alec's clear memory, mind for facts, and his way of speaking carried whatever he said as certainly as truth. Besides, Alec's papers were all in order and his legal documents—the patents leading up to the invention—were more thorough than Gray's. The other inventors didn't hold any real sway: Edison hadn't even designed his transmitter until after he'd seen Alec's telephone at the Centennial. Dolbear claimed to have had the idea of using undulatory current, but he hadn't filed any patents to the effect. Western Union's witnesses fumbled, but Alec took to the stand with confidence and majesty—he'd been trained, after all, in declamation. When he spoke he commanded all attention.

In the meantime, the public watched as this scrappy new company took on one of the biggest monopolies of the time—and was winning. Stock in the National Bell Telephone Company began to rise. Back in March, after Gray's cross-examination, stocks were $65 a share and Mabel begged Alec to sell off seven hundred of her stocks at that value, worried that the value wouldn't last. "Please sell out immediately," she wrote, "please, *please, please, please, please, PLEASE PLEASE PLEASE.*"

By summer, Mabel had sold off the shares, but still hung on to 810. In September, when Alec's testimony was ending, Alec sold two hundred of them at $300 a share. By October, they were selling for $500 a share—Alec sold fifty at that price and another fifty at $525 per share.

In the end, Western Union decided it was too risky to go on with the case. The final out-of-court deal was signed, in Alec's favor, on November 10, 1879. Western Union signed over all rights to telephony to the National Bell Telephone Company for the remaining years covered by the patent. In return, Bell Telephone agreed to stay out of telegraphy, and to pay Western Union 20 percent of all telephone receipts for the next seventeen years. They also agreed that Western Union would maintain their relationship with the Associated Press; news stories would continue to be sent via telegraph.

There was no official ruling on the matter, but it felt like a victory nonetheless. Gardiner wrote to Alec that "it is a very much better bargain for us than I expected, we could have hardly expected as much even if we had obtained an injunction against them." Western Union transferred all their existing telephones, lines, switchboards, and telephone patents to the National Bell Telephone Company, at cost. The stock prices hit $1,000 a share in the wake of the announcement.

Thomas Watson wrote, "I felt as if a crushing weight had rolled off of me." He caught the train to Marblehead and walked along the beach, shouting poetry to the sky.

By the New Year, there were over thirty thousand telephones in use and, despite his frustrations with the telephone, Alec found himself drawn again to tinkering. On February 15, Mabel gave birth to their second child, whom Alec wanted to name Electra or Photophone, after a new invention he was working on to send sound over a beam of light. But Mabel chose the name Marian, after her sister who died in infancy. Mabel checked with her mother to see if it would be okay, and Gertrude said yes, but so softly that Mabel decided to give the name a few days' trial. In the end, the child christened Marian would be known for the rest of her life as Daisy.

In August 1880, a month after he left the telephone company for good, Alec won the prestigious Volta Prize from the French government. It was prize of 50,000 francs, equivalent to $10,000, established by Napoleon Bonaparte in 1801. It was given in honor of a great electrical achievement—and only when something warranted extraordinary honor. Before Alec, it had been awarded only twice.

In September, Alec traveled to France to accept his award. As he crossed the Atlantic his mind was already wandering back to the deaf. With the telephone out of the way, and $10,000 in the bank, he wanted to begin a revolution. There was almost nothing that could to stop him. All was happy and well—or well enough.

Chapter 13

Nothing will ever comfort me for the loss, for I feel at heart that I was the cause.

—Alexander Graham Bell

In mid-June 1881, Alec and Mabel temporarily moved back to Cambridge, back into the Hubbards' old home to escape the stifling heat of a DC summer. In Cambridge, the sun shone strong, but was tempered by the breeze, "blowing fresh and cool through the trees." Even without Mabel's parents there, Cambridge felt more like home to them than their home in Washington did. Mabel found it "so lovely . . . and yet so sad." She loved to watch her children play, and she was now pregnant with her third child, but all the markings of her parents made her miss them. They were still in Washington, and she hated to be separated from them.

Mabel's own role as mother still felt foreign to her, and only more surreal for happening in the Cambridge home. "To think that I am 'Mama' now in the place where I was a little girl, that my babies run around where I ran, sit in the same high-chair at the table, and play about the same games as I did—I can hardly realize it all."

In Cambridge, the strawberries had grown as big as peaches, the grapes were ripening well, and the girls, too, were growing up. Elsie was "so very brown and rosy, a little round fat thing bright and happy," and Daisy was looking more and more like her father and developing a distinct fearlessness. Alec began to take Daisy on "expeditions" into Boston, just the two of them.

Alec had been sick all winter and was still weak, but he wasn't going to waste his time in Boston. Back in the center of oralist development, he met with Miss Fuller, whom he knew from his days at the Boston School for Deaf-Mutes, to learn more about her experiments in education. When Elsie had a party with eleven other children, Mabel could barely tell they were deaf. She didn't even use that word to describe them, saying only: "They made so few signs that I could not tell that they were not like other children." Now Alec was at work on a new organization, the American Association for the Promotion of Teaching of Speech to the Deaf, to bring together oralist teachers and share ideas, and he was still overseeing experiments in sound being pursued by his associates in DC. As June began to come to a close, Mabel wrote in exasperation to her mother, "Alec says he would rather die than leave work."

As Mabel and Alec approached their fourth wedding anniversary, Eliza wrote to them, "may each succeeding year bring with it a fresh stock of health, love, and happiness with as many new little loves as can be conveniently admitted to the home circle."

But on July 3, a little over a week before they would celebrate the date, the newspapers carried nation-shattering news to Bostonians: "A Dark Deed," began the sixteen-line headline of the *Boston Globe*, "President Garfield Assassinated / Shot in the Depot at Washington / By a Disappointed Office-Seeker, / When Starting on a Pleasure Tour / Internal Hemorrhage Sets In, / And the Physicians Give No Hope . . ."

The news hit the nation hard. President James Garfield was beloved across party lines. He was a charismatic speaker—even as a junior congressman, he was selected by his peers to give a eulogy for President Lincoln on the first anniversary of his death—but more than that, he was a man of great character. An outspoken abolitionist and the strongest friend the deaf had in Congress, Garfield always had his eye toward helping those whom others were likely to turn away from.

Now anyone who could possibly help turned their attention to the president. Alec was no exception. His mind did what it always did: found a singular focus and let the rest fall away. Right now it was the president, and more, the bullet still lodged inside him. Whatever the doctors said or the papers reported, he knew there could be no predictions about Garfield's life so long as the bullet remained hidden.

He wasn't the only one focused on the bullet; he read the accounts of doctors cutting into the wound, pushing their fingers deeper and deeper, probing around in the president's body looking for it. "Science," he wrote, "should be able to discover some less barbarous method."

Despite the fact that he'd moved his life to Cambridge for the summer, Alec hadn't left inventing behind. Back in DC, he had built a lab in a repurposed old stable that still stank of manure. The Volta Lab, as he named it, was where he and his team continued their scientific experiments with sound, focusing now on improving the newly invented phonographic record. His assistant, Charles Sumner Tainter, was working out of that lab on July 2, when he got the phone call that the president had been shot. Like so many Washington residents, Tainter left his work as soon as he heard the news. He joined a pilgrimage of sorts, walking dazed to the site of the assassination attempt, the Baltimore and Potomac train station.

Back in Cambridge, Alec was at work trying to find a way to shine light through the president's body to find the bullet.

"I cannot possibly persuade him to sit . . . ," wrote Mabel, "[as] he is hard at work day and night." But even as the work consumed him, he couldn't solve the problem of how to make a bright enough light without using tremendous heat. Then he read in the newspaper that Simon Newcomb, an astronomer and friend, thought that electricity or magnets might be able to locate the bullet. He immediately contacted Newcomb. A few days later, another scientist, George M. Hopkins, suggested the induction balance, an early metal detector that Alec had worked with.

Alec went back to William's shop and began work based on adapting an induction balance. A few weeks later, with Mabel late in her third pregnancy, Alec left for Washington with his device in tow.

On July 14, Alec arrived in Washington, a city swarmed with flies and smelling like a stable. He was met at the train depot by Professor Newcomb and the president's private secretary, who invited him into the president's carriage. They went at once to the White House, passing hundreds of people anxiously camping out on the White House lawns, waiting for news.

Inside, Alec met with doctors, discussed his invention, and gave a basket of grapes to the first lady, Lucretia, with a card: "To Mrs. Garfield, a slight token of sympathy from Mr. & Mrs. Alexander Graham Bell." Before he left, Alec received a card that allowed him access to the White House at any hour.

At the time, inventors tended to be secretive about the mechanics of their inventions, but in the race to save the president, competition was put aside. Alec reached out to anyone who might have answers, and everyone gave freely of their knowledge. By the time he reached his lab, Hopkins had already forwarded him a Hughes Induction Balance from New York. Alec and Charles Tainter set up the invention and, in a lab that sometimes reached one hundred degrees in the thick heat of that summer, they desperately began work to make the invention functional. Alec would listen at a telephone receiver and draw the electrically charged coils along a surface—Alec would hear a clicking sound when the electrical field responded to the presence of metal. The contraption worked fine for well-conducting metals, but bullets were made of lead, a poor conductor of electricity.

When Alec reached back out to Hopkins for suggestions, Hopkins began constructing his own induction balance in New York. Professors from Johns Hopkins telegrammed their suggestions. When Alec sent a telegram to one scientist to please send a workman to help, he wrote back that he was shorthanded but would come himself. On Friday morning, July 15, he arrived, stripped off his waistcoat, and got to work.

Before each test of their evolving design, Alec and Tainter fired a bullet against a board, flattening it out as it would be flattened inside the president. Then they would conceal it and have the other try to locate it with the device, taking turns holding a flattened bullet in their mouth or under an armpit. By July 16, they could detect a bullet in a clenched hand.

That day, Mabel sent Alec a package with a nightshirt, shirts, collars, cuffs, and handkerchiefs. Alec worried about Mabel, late into her pregnancy and now without him: "You poor little wifie, I feel so anxious about you. How do you manage about going up and down stairs?"

Mabel wrote back, describing the difficult scene at home. Her doctor didn't want her going out into the city, and Daisy, who was newly walking, was afraid of the new nurse she'd hired. Elsie was pouty and misbehaving, and when Mabel sent Daisy off to give her sister a kiss, "which the little thing solemnly toddled off to do," Elsie held her face high to reject it. Mabel sent Daisy to her sister again, and again it was no use. Finally Mabel gave Daisy a kiss and a flower, and Daisy took the flower to her sister as an offering. Again, Elsie turned away. Turning away from her sister was one thing, but Mabel worried that her daughter was turning away from her, too. Elsie didn't speak to Mabel, not nearly as much as she spoke to others; instead, she pointed at things to communicate.

"I fear she may learn to give her confidence and tell her little stories to others until it will be too late for her to care to come to me with them," wrote Mabel. "I know she is fonder of me than anyone else but she does not talk to me as she does to others—from what her nursemaid says, she must be full of childish prattle and pretty dictatorial ways—all of which I see nothing as she talks so indistinctly, hardly moving her lips."

It was obvious, at least to Mabel, that her elder daughter already saw her as different from other people. Mabel felt the pain of this realization acutely, but she was unsure how deeply that might change her relationship to Elsie, how much it might affect her ability to mother. She sat in her parents' home, considering her own mother, considering that distance between her and Elsie.

She wrote letters to Gertrude: "If I could be as much to my children as you have been to me my mother . . ."

Soon, Alec could detect a bullet in a bag of cotton, but the farthest distance the detection worked was two inches deep. They kept adjusting the invention. President Garfield hadn't strengthened, exactly—he was still too weak to sit up—but he had stabilized. Alec continued to work diligently, never forgetting that time could be up at any moment. He tried to locate a bullet in a bag stuffed with wet bran, or in a joint of meat he hoped would "more nearly approximate the dreadful reality." At the Volta Lab, Alec and Tainter worked late nights, experimenting and adjusting and experimenting again. They were trying to increase the distance that the induction balance might detect metal, but time and again, Alec's invention fell short. They altered the design, the battery power, anything they could think of, and managed to increase the distance from two inches to three. Soon after, Garfield's doctor, D. W. Bliss, visited the lab to observe the invention.

That night, Mabel had a dream that Alec tested his device on the president and it failed. She wrote him the next day. She wanted him home. "Surely Garfield is strong enough now to give away a little to your convenience," she wrote.

But earlier that day, unbeknownst to Mabel, the president took a turn for the worse. He was restless and exhausted, and his wound expelled an unusual amount of pus. The pus itself wasn't seen as a problem—at the time, it was common to believe that pus was a sign of health, and Bliss's medical bulletin that night reported that the wound "was looking very well," having "discharged several ounces of healthy pus." But at seven the next morning, Garfield was running a fever of 101. By 10:00 a.m., the fever hit 104, and he was vomiting bile.

The next day, July 22, Alec welcomed a Civil War veteran to his office. The man had lived with a bullet inside of him for over a decade; now

Alec would try to locate it. He ran his coils along the man's body several times, and had Tainter do the same. He could hear a small shift in sound at a single spot more than once, but it still seemed too faint. "I find that very feeble sounds like that heard are easily conjured up by imagination and expectancy," he wrote to Bliss.

At the White House the next evening, surgeons arrived for the president, operating to drain a pus sac three inches under the wound. That day's bulletin announced that "the President bore the operation well," but in truth he was still vomiting, still covered in sweat, the infection still spreading.

By July 24, Mabel hadn't heard from Alec in days. "I do hope you will be able to get some rest this summer, my poor little boy. Are you very tired out and does your heart trouble you much. I wish I could be with you and try to take care of you."

On the twenty-fifth, when the physicist Henry Rowland suggested the use of a condenser, Alec and Tainter managed to gain another half inch. Alec declared the device good enough to try, and told the doctors to send for him when they wanted to make the attempt. He believed it would be several days before he would be wanted, and went home, to bed. "I felt tired, ill, dispirited and headachy and . . . thoroughly exhausted from several days and nights of hard labour."

Mabel hoped the experiment would work so that Alec would return to her. She thought of Lucretia Garfield, her courage and strength, the way a whole nation "loans upon her courage." And Daisy, "full of life and mischief," and spunky Elsie, who wanted to kick the man who shot the president. The fact that her father was going to try to make the president well made no difference to her—she still wanted to kick Charles Guiteau.

Alec slept until eleven the next morning. Then, over breakfast, he received a letter from the White House. Alec was to report at 5:00 p.m., to use the device on the president at 6:00 p.m., when they planned to change the dressings on his wound.

Alec brought his invention through a back door to the White House and climbed a narrow servant's staircase. While Garfield slept, Alec surveyed his room to determine how to set up the induction balance, but he couldn't help but pause and linger on the president, remembering the way he had once looked, like "a man who indulged in good living—and who was accustomed to work in the open air."

Now that look was gone. "[Garfield] looked so calm and grand he reminded me of a Greek hero chiseled in marble," he wrote to Mabel afterward. The president lay on the bed with his head turned to the side, facing a screen. His eyes were closed, asleep, and despite it all, Alec thought he looked calm, even peaceful. "His face is very pale," he wrote, "or rather it is of an ashen gray colour which makes one feel for a moment that you are not looking upon a living man. It made my heart bleed to look at him and think of all he must have suffered to bring him to this."

As Alec and Tainter got to work setting up the invention in an adjacent room, the president woke and his doctors gathered to dress his wounds. For some reason Alec couldn't figure out, the induction balance kept making a sputtering sound, though it should have been silent. While Tainter adjusted the device's wires to try to solve the problem, the doctors beckoned Alec into the room. It was time. He would have to run the experiment without solving the problem.

Garfield watched with apprehension as wires were draped across his body, and asked Alec to explain the invention before they began. After Alec explained, a man knelt by Garfield's bedside. As they turned Garfield to his side, the president clasped the man by the neck and buried his head in his shoulder, only his eyes peeking out. No one spoke. They lifted his clothes from his body, exposing wounds down to his thigh, and Bliss ran the coils down Garfield's back.

Bliss operated the exploring arm while Alec listened at the telephone for the buzz that would indicate the location of the bullet. Garfield's eyes never left Alec, but from the telephone, Alec heard only the sputters. He did not hear the bullet.

Chapter 13

In Cambridge, Mabel was readying for what seemed like the imminent return of her husband, but Alec soon wrote that he'd be staying in Washington. Mabel was so upset that, in an attempt to make her feel better, a friend lit the letter on fire and danced a war dance before extinguishing it.

By then, Mabel was accustomed to her husband's absences—though no more pleased about them. They had begun almost as soon as their marriage, him splitting his time between London and the Greenock school, then between Washington and Boston to fight the court case. Back then, they swore to end the separations, but when Alec had a project, he still disappeared into it. Now Mabel waited at home, her belly growing with their third child, and weighed her desire for her husband against the life of her nation's leader.

It rained every day. Mabel busied Elsie with making bracelets from watermelon seeds but longed to see the sun. The sun, and her husband. Mabel wrote to him: "Dear love, . . . keep up your courage and do take care of yourself. I don't want you to be sick too. I'm afraid we neither of us would be the examples Mr. and Mrs. Garfield are!!"

By August 1, Alec had succeeded in locating a bullet in the body of a soldier, and they were to try again on the body of President Garfield. But again, Alec didn't hear what he expected; instead, there was a faint buzz over a wide area. The newspapers declared the experiments a wild success, but it wasn't true. They were no closer to finding the bullet.

Later, Alec would learn that the president's bed had not been stripped of metal, as he insisted it be. His hair mattress rested atop a spring mattress, a new and uncommon invention that no one had thought to consider. But further experiments showed that this wasn't the cause of the buzz.

Tainter kept working on it in Washington. But Alec could not put off his return to Cambridge any longer. Mabel had fallen suddenly ill. Alec returned to Massachusetts, but his mind was elsewhere; he still could not perfect a device to save Garfield's life. Within days, when Mabel seemed stronger, Alec returned to Williams's shop.

Back at the Volta Lab, Charles Tainter had successfully tested the induction balance's new attachment, a needle probe, but by then Garfield's health was too delicate. On September 5, about a month after Alec's departure, Garfield called Bliss in and told him he would be going to the sea; he would be going there to die. "I don't want any more delay," he said.

The next day, Garfield arrived in New Jersey, where he stayed in a bed that faced the ocean. On the night of September 19, Bliss sent out word to those in the house and nearby that death was imminent. All of those who had traveled to the sea with the president quickly and quietly gathered around him. They stood as only the sound of his rattling breath filled the room, and remained quiet after it stopped, after Bliss lifted his head from Garfield's chest and said, "It is over."

There was no crying, no sound, and then everyone but Lucretia left.

Lucretia stayed by her husband's side for another hour before someone guided her wordlessly away.

On the night of September 19, Alec heard the newsies cry, "Extra Republican—Death of General Garfield!"

"I hope indeed that there may be an immortality for that brave spirit," wrote Alec, who had always had a skeptical relationship with religion and God. "If prayers could avail to save the sick, surely the earnest, heartfelt cry of a whole nation to God would have availed in this case."

But in the Bell home, the tragedy of Garfield was a footnote to another grief. A little over a month earlier, on August 15, Mabel had gone into early labor and given birth to a premature son, Edward. Over the next three hours, they watched as he struggled to breathe. They watched as, for three hours, he died.

In the end, Mabel thought his face looked only like he was sleeping.

Daisy's first memory is hearing that angels had brought her a little brother but angels had taken him away again.

Mabel was slow to heal, pale and weak for months afterward, and her daughters caught her crying quietly to herself. She feared that she was being punished by God and called herself "only half a mother." For her son's death, Mabel blamed Garfield and Guiteau, and she blamed what she imagined as her maternal unfitness, which she connected to her deafness.

Alec blamed only himself.

"We smile at the eastern sailors who fall down on their knees before small wooden idols to save them from the storm," wrote Alec on the day Garfield died. "If the storm passes they believe their prayers effected the change and if it does not pass—why—it does not effect their belief in the least—it was the will of their God and they must resign themselves. We smile at them because we now know that the winds and the waves are controlled by immutable physical laws—and that not all the prayers and entreaties of the whole human race—could affect the movements of one drop of the ocean."

Chester Arthur was sworn in as president by 2:15 a.m., September 20, fewer than four hours after Garfield's death. Garfield's body was returned to Washington, and for two days and nights it lay in the Capitol rotunda as an estimated one hundred thousand mourners stood in line to pay their respects. The whole city, it seemed, was draped in black, in quiet.

Alec, too, was quiet. He may have won the telephone, but he couldn't save the president; he had Mabel, but Mabel was oscillating between grief, loneliness, restlessness. His only son was dead. And while Alec believed that all intelligent people believed in science over God, he let his wife have God.

He wrote to her: "I can appreciate the beauty of a trustful faith like yours—if I cannot follow you as far as you go—and I can join you in the earnest hope that it may all be true." He wanted her to share that faith with their children.

Elsie, in particular, who was then three years old, was having a hard time with the death of her brother. Someone had told her that he had been "shut up in a box" and "put in a hole in the ground." Alec was sure that it was eating away at her, and asked Mabel to talk with Elsie more about faith, "about a *good God* and *a hereafter* . . . lift her eyes to something higher and more beautiful . . ."

"For my own part," wrote Alec, "I would rather teach her what I believed to be a beautiful lie than a horrible truth."

Mabel believed Alec, too, was close to death. Not that he was ill, exactly, but that he was overworked to the point of collapse. The doctors agreed: he needed a good long rest. As a yellow fog set in around Cambridge, clinging to the particles of soot in the air, Mabel wrote a letter to Alec's parents.

"You know Alec has not been well for nearly a year," she wrote, "and the doctors say that entire rest from mental labor is necessary to avert serious constitutional trouble." They were planning an escape from work, a vacation to Europe.

She took breaks as she wrote, still weak from her labor and loss; the very yellow air she had to look through tired her. But she was worried about Alec: "I do really feel as if Alec's whole life's health, if not his life itself may depend on our going."

Alec made his own addition to the letter before it was sent: "Dear Mamma don't believe half my little girl has said. I am all right although I need rest, but a European tour is a thing Mabel and I long to make."

As they were planning their trip, President Garfield's autopsy was published, finding that the bullet had followed an entirely different trajectory than what the doctors had suspected and had lodged in a different part of the president's body, behind his pancreas. It was lodged too deeply to have been detected by Alec's apparatus—and it was nowhere near where the doctors had directed Alec to use the device. The bullet itself was harmless. Garfield died from the unrelenting probing of the wound by doctors' unwashed hands.

By late October, the Bells were traveling Europe together. "I will be your obedient and humble servant for the next few months . . . ," Alec wrote to Mabel, "any where in the wide world you wish I will go." In Europe, Alec became obsessed with caterpillars, bought two monkeys to have in their hotel rooms for fun, and carried Daisy around on his shoulders. Alec loved animals but never took responsibility for them. Mabel didn't like them, but they liked her. Still, she was gentle, reading and turning her pages with one hand so as not to disturb the monkey who had fallen asleep in the other, a look on her face one part helpless, one part disgusted, one part resigned.

But the joys of travel weren't enough to wash away the past, and even after they returned, in the summer of 1883, the death of their baby continued to resonate.

Small lessons for the children echoed with grief. After punishing Elsie, Mabel explained that she was withholding candy like God was withholding children from her. Mabel had asked for a child, she told Elsie, and she was to be careful with that child but she wasn't careful enough, and so God took the child from her. Elsie hugged her arms tight around her mother's neck.

When the girls were careless with their dolls, Alec was adamant that he and Mabel not replace the smashed dolls' heads. "Should not a doll be to a child as a child is to us?"

He thought the children should treat the dolls as they would one day treat their own children. "Let the dolls be given to them as children are supposed to be given to us," he wrote. "Let the dolls come with a delightful air of mystery about them. Where did they come from? 'God sent them'—or 'the angels brought them'—or—let Santa Claus bring them. We don't give them dolls—we don't buy dolls. They come—or are sent."

He hoped the children would grieve over the loss of a doll the way they would grieve over the loss of a child: "If our child should die from gross carelessness would not our grief be more bitter because we should blame ourselves for negligence . . . No other child could possibly replace the lost one."

He made no connection between how their children might grieve their dolls and how he and Mabel grieved their son, how he carried the guilt with him, and she with her. He did not think of how this grief would shape him.

Instead, he was impressing on his children how important it was to care for those around them. God sent the children, or angels did—it was their job only to care for and love the children brought, but Alec would soon go further than that. He would start playing God, trying to control birth, trying to bring about an end to deafness.

PART III

Chapter 14

A new epoch in the history of the mute world is dawning.

—Edward Hodgson

Alec was soaked through with grief. He'd lost the president of the United States, he'd lost his own son, and he was pulling away from Mabel, the most grounding force in his life. He was looking at the world through the lens of loss, and he returned to his work with the deaf with the ferocity of a man who was losing that battle, too. Oralism was spreading through the nation, but the manual method had a greater stronghold. Now Alec would focus his efforts keenly and directly. He'd put everything he had into winning.

The groundwork for this win was already in place, and nothing exemplified this more clearly than the Second International Congress on Education of the Deaf in 1880, or the Milan Conference, as it came to be known. Alec was receiving his Volta Prize from the French government at the time, but Edward Miner Gallaudet had attended, and he had been fuming about it ever since.

From the beginning, Gallaudet had had suspicions about the congress. It was supposed to be a convention of teachers—the congress itself was made up of teachers, mostly oralist—but instead everywhere there were priests in black cassocks. It wasn't that they weren't involved in deaf education, but their presence was so overwhelming that the lone deaf attendee, James Denison, believed for a moment that he was at the wrong conference: "Might I not be intruding . . . upon a solemn ecclesiastical convocation, discussing points of doctrine or church policy, concerning which laymen have neither voice nor vote?"

Before them all stood the president of the conference, Abbé Giulio Tarra, in his own religious dress, priests to his right and left, and beside them, a row of nuns in dark robes and white hoods. The Abbé Tarra presided over this conference so entirely that some suggested renaming the conference the In-Tarra-national Convention. He was described as "enthusiastic, magnetic, eloquent," but also as "constantly on his feet to plead for his favorite system; forever ready to interpose his official shield as the president of the Convention against all adverse thrusts." He guided the conference ever back to one issue.

At the outset, the congress's agenda was to examine a wide range of predetermined questions, everything from the age deaf people should enter school to whether they should sit or stand in the classroom. But despite the wide-ranging questions on the table, it wasn't long after the 9:30 a.m. commencement that it was determined that the congress would entirely skip the first two categories of questions—school buildings and instruction—and rush immediately into the question of methods: Should the deaf child be taught by speech or by signs? If they were taught speech, would they continue to speak after their education or would they sign? And if they did choose to sign—"if this evil does exist"—why? And how could they be stopped? Gallaudet was the second speaker of the day, arguing for the combined method—the use of both signs and speech education—but his speech was quickly dismissed. Not long after, his brother, Thomas Gallaudet, also spoke in defense of sign language. At the end of his speech, he paused. Then he lifted his arms and scooped the air in front of him before bringing his right hand to his forehead. "Our Father," he signed. As he continued the prayer, his hands were rhythmic and oceanic, moving like breath or song, sometimes soft around the edges, sometimes crisp in their definition. It was slow and mesmerizing; the sign for *forgive* was a repeated brushing of one hand against the other: "As we forgive those who trespass against us." He ended the prayer with a gesture everyone would recognize, two hands pressed together in a steeple, his head bowed in solemnity.

Before that moment of stillness could reach a natural conclusion, however, a voice cut through announcing the next event. Soon, more and more speakers were advocating against signed languages. Susanna Hull, that first teacher with whom Alec worked, detailed her experiences at the forefront of the oralist movement in London. As a young person, she had heard of Dr. Howe and Laura Bridgman, but it wasn't until 1863 that she first began to work with a girl who had lost her hearing. Since the girl had already begun to speak by the time she was deafened, Hull continued to develop her speech. Later, two congenitally deaf children came her way. She contacted Melville Bell and began the work of her life: "I had set before myself the goal of restoring my deaf children to home-life and society: what could more fully do this than vocal speech added to the language of books and writing?"

Back in 1868, Alec himself had traveled to London to teach Hull's students to speak, his first taste of the work. A few years later she visited the Clarke School to learn their methods. She was fully converted, and ever since learning how to teach lip-reading she had completely abandoned the use of either the manual alphabet or Visible Speech. She spoke against the combined system with an argument that would be used for decades to come:

"A 'Combined' system, in depriving the pupil of this required practice and constant care, injures the tone of the voice, and, as the deaf are only too ready to think themselves the objects of detractive remarks, persons so taught will soon find out that their speech is peculiar, and be driven to use their voices less, to depend on silent methods more, and to prefer the society of the deaf."

In other words, partial speech wouldn't be good enough. The only way speech would empower a deaf student was if it were developed to the highest possible quality. The pursuit of the perfection of speech began to eclipse the idea of speech for communication's sake. The goal was assimilation, passing. The deaf person should be able to disappear into a sea of the hearing.

To reach this goal, she believed that knowledge had to be deferred; it was "an end to be looked forward to," not an early pursuit of education. "We maintain that both teacher and pupil must fix a steadfast eye on spoken language as their single aim; that to introduce any other [studies] into the field till that has been acquired, is simply to impede the pupil's progress, by casting a stumbling block in his way." This was an important and dangerous shift from earlier oralist models, which still emphasized broader learning as a part of students' early educations.

To explain herself, Hull pointed to nature, to the fact that a hearing child learns language before she goes to school. "It would be the height of folly to propose to instruct an infant in physical laws, history, or grammar, the moment it commenced to utter sounds. . . . Why then must 'knowledge' be insisted on with the deaf, before correct speech has been acquired?"

Instead, her elementary school students worked on kindergarten-level tasks as they developed the ability to speak, as well as physical exercises intended to counteract the physical issues she falsely believed to extend from signing: "lung disease, distortion of the shoulders, [and] ungainly carriage."

Actual learning was delayed until a certain level of speech perfection was achieved. This perfection was always possible, she insisted. Any failures were merely the fault of the teacher. "I have found that with stronger faith in it, utter surrender of the mistaken desire for speedy knowledge, and more patient drill in the first elements of sound, failure cannot come."

Hull's speech was divided across the first two days of the congress, interrupted regularly for uproarious applause and, after the conference was over, would be printed and disseminated with congress materials.

In the days following Hull's speech, several advanced students demonstrated their abilities to the congress, sometimes through trips to the schools, and sometimes at the conference building. One British delegate,

Richard Elliot, was taken with them. He didn't necessarily believe they were better educated than their peers—"I should rather think the contrary the case"—but he *was* impressed with their speech: "They understand it better, [have] a readier means of communication, and, therefore, [have] a closer intimacy with the world around, in its every-day aspect."

The advanced students were followed by about thirty young men and women who had since left school and were living out in the world. Elliot didn't know if they'd been coached ahead of time, but either way, they came forward one at a time, answered the questions of the school director, and "spoke freely and fluently to the large audience." Though he couldn't understand the language—they spoke in Italian—he did observe that those who knew the language seemed to understand, "for there was every appropriate manifestation of attention on their part."

During the exhibitions, students used speech to the great pleasure of the oralists and, in Denison's words, "excited the admiration and wonder" of the nonprofessionals present as well. But many of the American delegates were suspicious.

Gallaudet was wary of the conspicuous lack of information about the students' backgrounds: Did they learn to speak before losing their hearing? If the students had only to retain speech, that could significantly impact their ability to speak afterward. Were they profoundly deaf, or did they maintain some of their sense of hearing? Students who had some amount of hearing also had a significant leg up in terms of learning to speak.

The students were almost *too* good—they appeared excessively studied, almost trained, for this moment. Denison expressed "positive lack of confidence whether everything was exactly what it seemed or was represented to be." In the schools the delegates visited, the teachers or the director performed the examinations of the students, and questions were limited to those things already taught.

And while certain students performed their ability to speak for the conference-goers, other pupils waited outside the classroom, comfortably combining gestures and speech to communicate with each other. As soon as they saw an onlooker, the signing abruptly stopped. When Denison asked them directly, in sign language, if they used signs, one group of these students looked at him with "a blank, mystified look," and shook their heads. But when Denison suggested that he'd already seen them signing, they "pleaded guilty, with a propitiatory smile."

Gallaudet soon abandoned hope for open debate. He saw that the people running the convention were "inflamed with the spirit of conquest, seeking for the present victory of a vote, rather than for the tardier triumph of truth that might follow unimpassioned discussion and a simple presentation of facts and results." The debate was over before it began. Gallaudet was right: sign languages didn't stand a chance in Milan.

He wasn't the only one who noticed. Oralists themselves were impressed at how quickly their cause took the fore. Hull wrote to Harriet Rogers, principal of the Clarke Institution, that "the victory for the cause of pure speech was in great measure gained, as many were heard to say afterwards, before the actual work of the congress began." Another oralist friend of Hull's claimed that it "was mostly a partisan gathering. The machinery to register its decrees on the lines desired by its promoters had evidently been prepared beforehand and to me it seemed that the main feature was enthusiasm and fervidly eloquent advocacy of the '*orale pure*' rather than calm deliberation on the advantages and disadvantages of methods."

At the end of the conference, the pupils were pronounced successes. It was declared that using both signs and speech was bad for speech, as well as for the precision of ideas, and it was concluded that "considering the incontestable superiority of speech over signs," the instruction of the deaf should proceed under pure oralist methods.

Sign language, in this official capacity, had lost the battle. The congress had no way of enforcing their resolutions, but those resolutions carried symbolic weight. It gave oralism, and oralists, clout and spirit; it was part convention, part rally. As the attendees returned home to France,

England, the United States, Scandinavia, Germany, Belgium, and Switzerland, they brought these ideas with them.

But the Milan Conference wasn't the only shift happening in the deaf world. The American deaf had had their own conference, the first National Conference of the Deaf, just days before Milan. Born of deaf collaboration, it was organized by three major deaf leaders: Edmund Booth, Robert McGregor, and E. A. Hodgson. And unlike preexisting deaf education conferences, it put deaf voices at the forefront. A total rejection of the idea of charity, it was a space where the deaf could speak for themselves on their own behalf, especially on the issue of oralism—both the way that it was harming the educations of deaf children, and because it was threatening the jobs of deaf teachers and principals.

It wasn't without conflict in its organization, but calls for peace and unity were constant in both the run-up and the conference itself. As it began, E. A. Hodgson, the editor in chief of the *Deaf-Mutes' Journal*, wrote that "the excitement has been intense, which is a sure proof that great hopes are centered in the great gathering. . . . A new epoch in the history of the mute world is dawning."

On August 24, 1880, hundreds of deaf people from across the nation made their way to Cincinnati, where they began to fill halls and corridors of the Gibson Hotel, where many were staying. It was hot, oppressively hot, but no one cared: "The interest and importance of the occasion compensated us for whatever discomfort we suffered," wrote one attendee.

They knew each other's names, through letters and newspaper columns and the stories of those they had in common, though they had never seen each other's faces. Now they found each other, whom they considered the "wisest and best" of the community.

Later stories would summarize what people were like in person for those who couldn't make it. Of the newspaper men: "Mr. Read . . . was the funniest; Mr. Hodgson, . . . the most reserved; Mr. George . . . the most determined and thoughtful looking; Mr. Gallagher, . . . tallest of 'em all, and Mr. Hays, . . . the most refined in manners."

As those with stature found each other in the crowd, three young men snuck out for their first cigar, and another wrote a poem about them:

Ah, bright their boyish fancies,
Wrapped up in wreaths of blue,
Their eyes grew dim, their heads were light:
The Gibson 'round them flew.

Attendees caught up on the news of the deaf throughout the country and shared stories of the strange hearing people they met in their travels there. One group had the dubious privilege of running into a prominent man who had "taken great interest" in the deaf and asked them, as though it were their first day being deaf: "Did you ever try to learn to *hear* by the motion of the lips?"

"[We] were astonished by his rare intelligence," one remembered. "After a little, we told him that we never thought of such a thing, but that we did at once attempt to understand the mysteries of lip-reading."

At the conference, there was no need to put up with such nonsense as that—the value of their own life experiences was not underestimated, and they were free to communicate in a way that they all understood. They could simply smile at the absurdity of hearing people who tiptoed down the halls. "Evidently," wrote one observer, "they were fearful of disturbing the proceedings."

On the first day, August 25, they gathered in the Bellevue House for the beginnings of discussion. "Among the audience not a word from their lips rent the air, yet everywhere was the liveliest interest displayed in the mute language of hands and eyes—a language which may appear amusing and meaningless to an ordinary observer, but which is really as rich in expression and beauty to those well versed in it as the English language itself," recalled a reporter who attended the event. It was a space, in other words, centered around signing deaf people.

Early on in the conference, the attendees named Edmund Booth the chairman of the first session. Seventy years old by then, Booth had been educated by Laurent Clerc, who had started the American Asylum with Thomas H. Gallaudet. He was also a successful newspaper editor in Iowa who had long argued for deaf independence, education, and access to ASL. Now he was escorted to the stage. With his long forehead, prominent brow, and full beard, he was the picture of a respected elder.

"*Fellow graduates,*" he began, "you are now assembled in the Convention, the first National Convention of Deaf-Mutes in the world's history. Your presence here is an attestation of the value of education, and this is further shown by the general intelligence manifested in your faces and flying fingers." In other words: intelligence manifested in ASL, a language he knew to be under attack. In just a few days, the Milan Conference would issue its tragic declaration calling for an end to combinist teaching—but the unity that this gathering represented would ensure ASL's survival.

Even the writing that emerged from the gathering is distinct from that of hearing-led conferences about the deaf—it carries a tone of celebration, liberation, unity. And not a hint of charity. In fact, the opposite of charity: over the course of this conference, the National Association of the Deaf (NAD) was organized to ensure these conferences would continue into the future, enabling a space where the deaf could meet to discuss and organize around the issues facing them as a people. The presidency was offered to Booth, who turned it down in favor of it going to someone younger. And so Robert McGregor, the founder and principal of the Cincinnati Day School for the Deaf, was appointed the first president; later, E. A. Hodgson would become its second. This association, grown from the collaboration of Booth, McGregor, and Hodgson, as well as from grassroots deaf organizing, would become a channel through which the work of deaf liberation could continue for generations.

Wrote one regular columnist for the *Deaf-Mutes' Journal*: "The absurdities of the past—that [deaf-mutes] live on charity, etc, are fast vanishing away, like the dew before the morning sun; the star of intelligence has risen and is shedding a light that puts to shame all quackish incongruities and groundless superstitions. The day is not far distant when mutes will be regarded as on an equal footing with their hearing brethren, and it is hoped that intelligent beings will use, and not abuse, their God-given faculties, unrestrained by the travels of a by-gone dogmatism—that deadly foe which clogs the wheels of thought and progress, and hinders the freedom and welfare of all mankind."

This convention, like Milan, had escaped Alec's notice. A month after the conventions, back before Garfield's assassination, Alec had written to Gallaudet with no awareness of these new turns of events. He spoke from the fixed point of what he already knew, and the isolated point of his own developing research. He wanted Gallaudet's response on two new papers he was working on: "Fallacies Concerning the Deaf," which focused on prejudices against the deaf, and another, about the inheritance of deafness, which he planned to present to the National Academy of Sciences.

Alec had also turned back to his first student in America, George Sanders, whose education had lapsed as Alec had gone on to bigger things. Now Alec began to work with him again on speech but hoped Gallaudet could help him find a more general teacher, perhaps someone from the college who could communicate with George through the manual alphabet. He wasn't opposed to this, though he still wanted to keep George as much in the hearing world as possible. "[I] do not wish to help him learn signs or to make many deaf friends," wrote Alec.

Alec believed that the deaf community was easier and more comfortable for many deaf people, but he didn't think it was better for them in the long run. He was diving back into the work, and back into the movement, but he had seemingly no idea of the momentous changes that were happening in the field, or how it would impact the reception of his work.

While Alec was blithely unaware of the oralist victory, news of the conference made its way through the deaf community, trickling down from the *Deaf-Mutes' Journal.* There, the community could read excerpts of Thomas Gallaudet's letters home from Milan, as well as coverage from New York, Hartford, and London newspapers. E. A. Hodgson provided no editorializing on this coverage: Thomas Gallaudet's letters were more or less neutral, reprints from the hearing press were generally favorable to oralism, and E. M. Gallaudet's observations flashed with fury at the partisan nature of the proceedings.

Instead, commentary was provided by regular *Deaf-Mutes' Journal* columnist "Mr. Why," who quoted excerpts from E. M. Gallaudet's Milan coverage printed in the *American Annals of the Deaf and Dumb*, bringing attention to a particularly upsetting anecdote of a deaf pupil who said that "a deaf-mute without speech was no more than a brute, an ape."

"What must be thought of a system of instruction that fosters and applauds such a degrading and brutal sentiment?" writes Why. He says the pupil was not to blame for this belief, that Why himself had met pupils of the oral method who had once shared in that sentiment: "They admitted that while at school they were taught to look with horror upon a mute who used signs, as if they were the most degraded of beings."

Pointing out that most orally taught deaf people could only speak well enough to be understood by their families, he writes that it is ridiculous to see them "affecting to be far above a mute who may be vastly superior in the command of language and general information, even if he does use signs." But this was part and parcel of oralism: Devalue sign languages. Devalue those who used them.

Meanwhile, George Sanders, who was roughly fourteen years old, was taking classes at the Kendall School, the primary and secondary school attached to the National Deaf-Mute College. His parents had enrolled him, but now that Alec was becoming more involved in George's education again, he wanted to enroll him in a hearing school. James Denison, George's teacher and the one deaf person present at the Milan Conference, wrote to Alec to beg him to reconsider.

Denison had observed George's language abilities to be advanced and impressive, but he had concerns about George's more general education to the degree that he had considered breaking him off from the rest of the class in order to give him special lessons. "He needs much more practice in simple mental operations," observed Denison.

"It seems to me," he wrote, "that this backwardness in Arithmetic has mainly arisen from his lessons in that study having been given too much and too entirely orally, and without sufficient repetition to fix principles in his mind." Denison, as a deaf person, could see into George's experience in a way that Alec couldn't, and tried to elucidate it for him. "It requires a certain amount of effort and concentration of mind for him to read from the lips, and this, I am sure, interferes more or less with his readiness and detracts from his ability to conceive what is expected of him." In fact, George had told Denison that he often didn't understand the questions being asked or the answers being given.

"How important as it is for him to improve in articulation it certainly can not be necessary that his progress in Arithmetic should be impeded on that account." He suggested that if he was going to be in a hearing school, the teacher should write even the shortest questions down in writing so that he could read along to whatever spoken language was being used—that way the guesswork of lip-reading would, at least, be eliminated.

"He has proved himself an amiable and docile pupil, and has seemed happy at Kendall Green, and I am, of course, sorry to part with him," continued Denison, cordially. "But I know you have his good at heart and that the decision you have reached as regard to his future has its points of recommendation." And then, in the tiny space between that sentence and his signature, Denison crammed in one last addition, too important to be omitted. He turned the final period to a comma and, in two lines of tiny print, continued the thought to its rightful end: "Though to my mind, the plan of sending him to an ordinary school has serious disadvantages."

But Alec still didn't question oralism, even as it had obviously failed one of his first and most beloved students. He couldn't believe the failure. He believed, instead, that there was rampant prejudice against the oralists and their cause.

Several years after the Milan Conference, as the deaf were taking the earliest steps toward the organization of their own international conference, Alec was working to begin an organization in which oralists could speak freely. By then, he believed that oralist teachers were being pushed out of the broader conversation about deaf education and needed an opportunity to organize on their own. Sarah Fuller, the first teacher he'd worked with in America, bemoaned that Gallaudet's organization and the *American Annals of the Deaf and Dumb* still had total control of the conversation, and said that Alec's were some of the only words that would penetrate it.

"We need your words," wrote Miss Fuller, "for the other side—the advocates of signs—have had about a monopoly of the magazine. Perhaps I ought not to speak so freely of the matter, but there are very few persons to whom I can say anything about these matters." Miss Fuller, along with other oralists in Alec's community, helped inculcate him in the belief that they were the oppressed group in the fight over methods. Their philosophy, they believed, was subject to disproportionate criticism and was all but blocked from publication in the most prestigious scholarly journal pertaining to the deaf. Understanding nothing about why that might be—since he had no awareness of the silencing that had occurred at Milan or the political organizing in Cincinnati—he moved forward aggressively. He believed his philosophy of oralism, and its supporters, were being wronged. Now he would right the way by any means necessary.

The first part of his initiative seemed harmless enough: on October 1, 1883, Alec opened his own kindergarten, with a school for hearing children on the first floor and a school for deaf children on the second. Surrounded by stately Washington, DC, homes, the school was set back from the street, old-fashioned and vine-covered.

Alec was as well-known as ever—he had secured telephone fame, been made an officer of the French Legion of Honour, and attempted to save the life of the president. In 1882, the same year Alec officially became a US citizen, he and Gardiner began a new collaboration to save the floundering *Science* magazine, and he and Mabel moved into what was speculated to be one of the biggest and most expensive homes in DC.

What he wanted most, though, was a return to peace, to the work he loved and believed in. This school would be Alec's kingdom—if oralism's effectiveness still needed to be proven, then he would prove it here.

The school building wasn't finished yet, but the classes began anyway, the lawn scattered with integrated groups of children at play. It wasn't always easy, and the deaf children were shy at first, not knowing the rules of the games or how to communicate, but soon the interaction between the groups grew.

The way Alec saw it, there were two basic ways deaf children were being educated in America at the time: they were either put in schools with hearing children, and "the result of this has always been that one class of children has been neglected for the other, and justice done to neither," or segregated into institutions of their own, "and this has had the unfortunate tendency of keeping them separate and apart from the rest of the world."

But the bigger tragedy was how many deaf children weren't being educated at all. Alec claimed that there were an estimated seventeen thousand school-aged deaf people (between the ages of five and twenty-one) in the United States. Of those, only seven thousand were enrolled in schools. While some may have already graduated, Alec still believed that there were at least as many school-aged deaf people not receiving an education as those receiving one. He set out to transform this, to open small oralist day schools across the nation to encourage more parents to enroll their deaf children, without having to worry about sending their children far away to residential schools, where the children would learn a language the parents didn't share.

In this model school, the educations of the deaf and hearing children would be separate, but social time would be shared. It would be a constant interplay of learning and application. The deaf children would learn from the hearing children what speech was, how to read lips, and why they might want to do either.

This wasn't exactly what had been proposed at Milan, but it was still in line with the advancement of oralism. Whereas the congress was more focused on speech at any cost, Alec wanted to prioritize learning, and wanted to develop speech by giving a real-world motivation to practice it. The classroom would teach speech, and the playground would inspire it.

Alec worked only with the deaf students, and gave them cards with the words for objects written out on them; they had to match the words to the corresponding items. A few of the students knew what words corresponded with the miniature soldier and tent, and so they started with those. Alec scrambled the cards on the low table that the school used in place of the usual bolted-down desks. He gave the teacher, Miss Manner, an object and asked her to find the word. She pretended to be perplexed by which card to choose, which thrilled the children, who scoured the words along with their teacher.

Day in and day out, in that room with morning and afternoon sun, a bay window overlooking the garden, and an open fireplace, the children played their way through their educations. Curtained shelves were lined with toys: a horse with real hair, steam cars, a doll with her own trunk of clothes. "'Play' is Nature's method of educating a child," wrote Alec. "If you choose to substitute for 'Nature'—'an intelligent mind'—then it is God's method of education!"

In early October, one of the new students, George, took the role of teacher for the day, instructing the students—Miss Manner and Alec among them—in a game they were familiar with, Sit or Stand, in which the teacher wrote an action on the board and students had to do it. Now with great seriousness, George took on the authority of a teacher, saying the word after pointing to it—or maybe not quite saying a word, but making a sound he imagined to be a word. Alec noted, "In giving these directions he imitated Ms. Manner exactly."

Later that day Alec began to teach him to write, the same way he had taught George Sanders all those years before. Alec took the words they'd been learning, the words George pointed to earlier, and wrote them lightly with charcoal on the board. When the word was written, George traced it over with his own charcoal. He had a knack for recognizing the shapes of words. He did well with the game of matching cards, and now he traced the word *sit* with little error. It seemed that language was beginning to develop. He was only mimicking now, tracing words, but Alec's hope was that that he would soon be able to communicate his own original ideas though English—and that he would do it without ASL.

Daisy and Elsie attended the deaf school, too, speaking clearly and distinctly in the way they learned to do with their mother. Alec hoped that the students would learn to read lips from watching the girls. Elsie, at age five, was learning to read lips. During the sit-or-stand game, Miss Manner mouthed directions without voice and Elsie made no mistakes. But Elsie already knew English; for the other students, reading lips meant accessing unknown sounds, unknown meanings. For them, it wasn't so simple.

This school was Alec's dream, the culmination of years of expertise in service of this larger goal of bringing the deaf and the hearing closer together—or bringing the deaf closer to the hearing at least. Now he was working toward this not only with speech but also with proximity. Bringing the deaf and the hearing close to each other wasn't only a theoretical part of the education, it was built into the very structure of the building, the very structure of the school days.

But already there were cracks in the facade. Mabel didn't want her daughters to keep attending the school. Elsie wasn't just lip-reading; she had started signing—both with the deaf students and with Mabel, who wrote that it was "very disagreeable and bad for the child." It was also an echo of what deaf educators long claimed happened in oralist schools: despite the prohibition, the children learned or developed signs on their own. They seemed to *need* this; it came forth from within, no matter the prohibitions. It happened in Northampton, it happened in Milan, and now it was happening in Washington, at Bell's own utopian school. But oralists didn't ask *why* sign language kept coming back; instead, they redoubled their efforts to stamp it out.

Chapter 15

[The students] not only have mouths to be taught to speak and eyes to be taught to hear—but they have something of far higher importance—the cultivation of which has been cruelly and sadly neglected. They have minds to think, ideas to express—questions to ask.

—Alexander Graham Bell

The school was a microcosm of the first part of Alec's vision—peace and interaction between the hearing and the deaf—but a second part of his vision was only just emerging. If the school could demonstrate the feasibility of oralism, he still needed something to show its urgency.

In September 1883, Alec was finishing his paper for the National Academy of Sciences, *Memoir upon the Formation of a Deaf Variety of the Human Race*, the one he'd written about to Gallaudet in the wake of Milan. Back when he began the study, in response to a request from the Massachusetts State Board of Health, he'd sent questionnaires to several deaf schools to gather statistics on inherited traits. As was his wont, he'd been working tirelessly on the study, pulling together the data points and considering their implications. Now he sat down with the results, lined them up, measured them. At first, he noticed names—not just common names, but names like Blizzard and Brasher and Gortschalg—showing up at the schools generation after generation.

As he had anticipated, Alec had found that deaf-deaf marriages more often led to deaf children, but this wasn't particularly surprising. The phenomenon of hereditary deafness had been widely known—articles on the issue had been published in the *American Annals of the Deaf and Dumb* for decades before Alec even came onto the scene.

What did surprise Alec was that it wasn't just a small percentage jump. When he aligned the procreation rates he had collected so far with preexisting hearing procreation rates, he found that according to his numbers the deaf population was growing faster than the hearing population. "*The indications are that the congenital deaf-mutes of the country are increasing at a greater rate than the population at large.*" Left unchecked, he thought, the deaf population would eventually subsume the hearing.

Twenty-four years earlier, Darwin had published *On the Origin of Species*, which overturned the prevailing worldview with his ideas on genetics and evolution. But the deaf population wasn't growing in line with Darwin's principles—it couldn't be the survival of the "fittest," Alec thought. Something else was at play.

By the early 1880s, Darwin's ideas were expanding: while natural selection ought to have ruled, human compassion was complicating the outcome of our species. In Darwin's words: "We build asylums for the imbecile, the maimed, and the sick; we institute poor-laws; and our medical men exert their utmost skill to save the life of every one to the last moment." This was resulting in "the weak members of civilized societies propogat[ing] their kind." The social safety net of mankind was undermining the laws of evolution: "No one who has attended to the breeding of domestic animals will doubt that this must be highly injurious to the race of man."

This was the seed of eugenics. There was something about humans that was resisting nature, and this was "injurious" to the human race.

But Darwin took it further; he argued that this impulse to care for each other was driven by the instinct of sympathy, which was a fundamental part of our humanity. We could not "check our sympathy, even at the urging of hard reason, without deterioration in the noblest part of our nature."

The sentiment that sympathy was central to the nobility of our nature wasn't shared. Darwin's cousin, Sir Francis Galton, coined the term *eugenics* in 1883, the same year as the most damning phase of Alec's career began.

Alec still believed in oralism. And if the first obstacle preventing the spread of oralism was that it was believed impossible, then the second great obstacle was that it was believed unnecessary. Alec hadn't explicitly set out to find a need, but once he stumbled on it, everything clicked into place.

The isolation of the deaf—an isolation he believed to be sometimes a result of prejudice, but more often a result of language barriers—meant that the deaf learned to associate only with each other, become friends with each other, and fall in love with each other. The cause, Alec believed, was education. Without the educational preparation to interact with hearing people, both linguistically and socially, the deaf were inadvertently making a "deaf race." Education was still, for him, about social integration, but these findings meant that the stakes were much higher than he once thought. His goal began to morph: now oralism wasn't only about the ability of the deaf to interact in the hearing world so that they would be empowered. It was bigger than the lives of individuals. The deaf needed to be prepared to assimilate into the hearing world so that they would be able to marry hearing people. Oralism was the key to minimizing procreation among the deaf.

Alec's harnessing of a eugenic-based fear of difference would give oralism the great last push that it needed to become the standard in deaf education. He insisted that oralism was urgently important—not just for deaf individuals but also for the human race.

When Alec first wrote to Gallaudet about these ideas, he was clear on the fact that he would not pursue legislative interference in marriages of the deaf, but his rationale was coldly pragmatic: "I hardly think that legislation concerning the marriage of the deaf would prevent the present condition of affairs, for, so long as deaf persons of both sexes continue to associate together in larger numbers in adult life I fear that the suppression of marriage would only lead to immorality." The way to change deaf intermarriage was to eliminate ASL from the classroom, because doing so would dismantle the community that kept deaf people together. As long as sign language–based education remained, the deaf would find comfort in ASL, would fall in love with people who used it. A law would criminalize their love but not prevent it.

In late October 1883, after a meeting with Chief Justice Morrison Waite, Alec began to second-guess himself—not on the ethics of his ideas, but on how far he was willing to take them. Waite suggested that a law would be necessary, and he talked Alec into agreeing. That night Alec told Mabel that he would add two legal proposals to his paper: a law "forbidding marriage between two congenitally deaf people with deaf relatives" and another "forbidding the assembling of large numbers of deaf mute children together for the purposes of instruction."

That night, after he told Mabel about his decision, he said he was tired and that "his heart troubled him so much." He went to bed at eight thirty.

Before Mabel joined him, she went to look one last time at her own children. She was pregnant again, and her girls were asleep in each other's arms, holding hands. Daisy's head rested on Elsie's shoulder, like a "picture for an artist to paint."

As Alec readied to present his paper, his children caught colds. Mabel woke up in the middle of the night, night after night, to go check on them. Soon Mabel, too, caught a cold, and her first cough in years. She was worried about being sick during her pregnancy and sent for the doctor, who told her to stay away from draughts. And so she went on, running the household and renovations, caring for her sick children, waking throughout the night to check on them. Alec traveled to Connecticut for the National Academy of Sciences meeting at Yale, where he was slated to present his findings.

In New Haven, on November 13, after opening remarks and before lunch, to an audience of more than one hundred people, Alec presented his ideas: that deaf intermarriage would eventually lead to "a vigorous but defective variety of the race," and that because of sign language–based residential education, that was exactly what was happening.

He said that more research was needed—that his study was merely a beginning—but he believed it was a solid foundation. He believed that the human race was facing an emergency. A Yale professor drew the connection to criminal asylums, where poor people met poor people. Another person asked about consanguineous marriages, or marriages between relatives. This was commonly believed to be a predictor of deafness, but Alec hadn't found consistent information on that. He held only to his original claim: more research was necessary, but based on his evidence, deaf people would form a distinct race if left unchecked. He believed that widespread oralist education could prevent disaster: it would decrease the use of ASL and increase ease with English, which would prepare the deaf to integrate into the hearing world, thus decreasing deaf intermarriage.

But he had changed his mind about legislation. He advocated against it. First, there was the possibility that such a law wouldn't actually stop the relationships themselves, leading to "immorality" among the deaf. Second, it would be difficult to determine whether a person was deaf from birth. This was essential, because he believed only those who were congenitally deaf had the possibility of carrying deafness—deafness caused by illness had no bearing on offspring. And finally, his analysis found that those with deafness in their family were those most likely to pass it on to their children, even if those people were hearing. The only sensible law would apply not only to the deaf but to the hearing as well: a law that prevented the marriage of two people—hearing or deaf—who both had multiple instances of deafness in their families.

But even that, he said, would require much more research and data before it could ethically be considered. His study was a beginning, he insisted. It was not enough evidence on its own. Instead, he used it to argue for oralism as a supposedly benign solution to a problem he was convincing the world was real.

At home, Mabel went on. She ran the house, stayed away from draughts, and checked on the children. And then one morning she awoke, shivering and feverish, by Elsie's kiss. The doctor told her to stay in bed but said there was no concern for the baby.

Mabel's oldest sister came to stay and spent the next night on the sofa. During the night they contacted the doctor again, and then, after the doctor still didn't show any concern, Mabel reached out to her neighbor to see if she knew of another doctor. She didn't, but came over to help watch over Mabel.

The night of November 18, as Alec traveled home from the conference, Mabel awoke to contractions. With the help of only her sister and her neighbor, she gave birth to a son, Robert.

"It was so pretty and struggled so hard to live," she wrote, "opened his eyes once or twice to the world and then passed away."

Alec arrived home only three hours later.

The assumption of Alec's *Memoir upon the Formation of a Deaf Variety of the Human Race*—that "the production of a defective race of human beings would be a great calamity to the world"—was galling to a group that had begun to consider itself an independent culture. Deaf people had their own social organizations, newspapers, churches, political groups, and far more, with deaf people in charge.

In the wake of the 1880 conferences, the signing deaf were beginning to solidify a story about themselves. In deaf newspapers, they cataloged and disseminated information about their history, and wrote the story of their liberation in a way that wasn't dependent on hearing savior narratives. One reprinted lecture began with the bleak history of deaf education—outlining how deaf children in France were considered a disgrace; that under Justinian law in the Byzantine-era Roman empire, they had no civil rights; that even St. Augustine believed that the deaf had no place in heaven. Often, they were simply killed.

But this writer cataloged how the world changed, too, one educated deaf person at a time. First there was Bede the Venerable, who made record of a deaf person being educated in 685 BC; by 1545, Jerome Cardan of Pavia declared that "the instruction of the deaf is difficult, but not impossible"; and there were accounts from the seventeenth century, including those of George Dalgarno, on which Alec relied in his earliest teaching. By the mid-eighteenth century, Abbé de l'Épée founded the first school in France, and Thomas Braidwood founded one in Scotland. This is where the continuous story begins: l'Épée and sign language versus Braidwood and speech.

The narrative wrests back ownership of the educational and social advancement that came with the signed method: "Our modern language of signs was invented by the pupils, not by the teachers. The children brought their own natural signs to school, and all the teachers could do was to improve and systemize them."

What was happening in the deaf world was a reclamation, a rewriting of history. The dominant narrative was that the deaf were a dependent class, but the deaf were teaching a different story. The narrative of dependence, of charity, was being overturned.

Edmund Booth argued that charity did their people harm: "On this idea all appeals were made for assistance—state, national, and private. Governments and individuals who gave were impressed with that single idea," he wrote. "In that age men had not come to understand that what they called charity should be regarded as a matter of justice." In other words, deaf education was not a gift but a right.

The deaf were a people empowered. They had found that empowerment within themselves, within their community. They were beginning to demand their due.

And they were pooling funds for a memorial of their most fundamental story, that of Thomas Hopkins Gallaudet's early work to enable deaf education. Though he had died in 1851 of dysentery, his one hundredth birthday would be in 1887, and they were planning to erect a statue of him to honor the origin story of their community.

They did not believe, as Alec said, that they were "defective," nor that their intermarriage was an "evil." Alec had already threatened deaf people with his promotion of oralism, which impeded the formation of deaf communities. Alec's study now took it further, proposing to dismantle the community altogether by controlling birth.

Alec kept pushing forward. In 1884, he presented his findings again, this time to the American Association for the Advancement of Science, and subsequently sent copies of this speech to all members of Congress, as well as deaf educators.

Early on, the attacks against his paper from within the deaf community stemmed largely from a sense that it was neither original, nor well-researched. The editor of *The American Annals of the Deaf and Dumb*, Edward Allen Fay, wrote, "Even if this theory were fully established, we should not regard it as a reason for abolishing our institutions for the deaf, since the good they do far outweighs the possible evil." Fay did, however, go on to say that the deaf should be warned of the possible outcomes of their marriages.

E. A. Hodgson, in the *Deaf-Mutes' Journal,* noted not only the long history of educators of the deaf discussing heredity but also the failure of decades-long attempts at discouraging intermarriage: "We believe it has for a long time been the custom of those eminent in the instruction of deaf-mutes, to advise their pupils not to marry others similarly afflicted, but this caution has been given in vain in the majority of cases, as human nature and the sympathy and congeniality of companionship which a common affliction and a common language inspire, have proven too strong to yield to the dictates of wisdom and experience." In other words, this whole debate was pointless. The deaf would marry whomever they wanted. Companionship trumped argument.

But soon, the whole debate would be colored by a false claim—a false claim that was coldly foreshadowing.

On December 31, 1884, the *New York Times* published an article with the headline "A Deaf-Mute Community: Prof. Bell Suggests Legislation by Congress." In its excerpts from Alec's *Memoir upon the Formation of a Deaf Variety of the Human Race*, the *Times* quoted from his discussion of legislation, including that "due consideration of all the objections renders it doubtful whether legislative interference with the marriage of the deaf would be advisable." Alec didn't want legislation, but the headline suggested otherwise.

On the same day, Washington, DC's *Evening Star* took the error a step further, writing that Alec advised "legislation be had restricting in a measure the intermarriage of deaf mutes." The rumor began to spread that Alec was petitioning Congress for a law forbidding the intermarriage of the deaf, setting off a whole new wave of resistance.

An article soon ran in the *Washington Post*: "The New York deaf mutes are very angry at the language of Prof. Bell. The mute press has criticized him severely for his bigotry." The piece went on to reference the phantom law, noting that "such a law, if passed, would be unconstitutional and wholly inoperative in practice."

If Alec was following these responses, it didn't show—not immediately anyway. He was absorbed by the loss of his son. It was the second time he'd experienced the birth and death of a child, the first only two years earlier. He was devastated by the loss, and worse, devastated by his inability to console his wife, who was busy consoling him.

Mabel blamed herself for the loss of her child: "Perhaps I got my strength at my baby's expense," she wrote. "The doctor said it was the fever that loosened my baby's hold."

Alec believed his own "ignorance and selfishness" were to blame, that it was his fault for impregnating Mabel again too soon after her first lost child. He began to pull away from her again. She was still weak after the loss of their second son, and Alec worried about being intimate with her. He worried for her life.

This time there would be no European vacation, no period devoted to healing. Within months, Alec had turned his attention back to oralism and to his prime demonstration of its feasibility: his school. He distracted himself from his grief, pouring himself into the work.

Alec loved his school. When Mabel made some comment about its expense, he said she could stop any other expenditure of his—"anything everything"—but not his school. Mabel learned the school was the line that could not be crossed: "to touch that dear little school was to stab him to the heart."

But already there were problems. Miss Hitz, Alec's hand-chosen head teacher, had left at the end of the previous school year to be married. It being an experimental school, defined by trying new educational methods and then measuring their results, it was hard to find a teacher both skilled and open, and the loss of Miss Hitz was a blow. Alec scrambled to find her replacement, Miss Littlefield, just before the start of the second year.

The second year of Alec's school opened on October 1, 1884. Four deaf children were enrolled, and the class continued to spend time with the hearing children. Downstairs in the hearing school they wove paper mats, stitched with colorful threads. They played a game where they stood in a circle with one student in the middle and sang, "Oh, look at little James, who shows us the game. Oh, look at little James, and we'll do the same." Whatever the child in the middle did, the rest followed. The deaf children joined in—George imitated rowing a boat, Gracie kissed her hand, and Floyd clapped.

The deaf students gathered upstairs, making faces into hand mirrors, learning about the muscles of their faces, their tongues. When they forgot the meaning of a sentence, they learned the word *forgot*. When one student was emphasizing it wrong—FORgot—Miss Littlefield took his hand and shook it once for each syllable—*for-got*—and then lighter for the first syllable and heavier for the second: forGOT.

At the school, they taught no signs, but they did learn the signs that the children brought to the school. Mostly, these seemed to the teachers to be home signs, and they worked to replace them with English. Each of the youngest students had a card rack—a board with several brass clamps to hold vocabulary cards, which had the names of objects written in both script and line-writing. Using these cards, they began to use English words to communicate their desires and ideas.

Alec taught them, too, when he could. He came in for weekly speech lessons, and on October 24, he took the students, now six of them, and one student teacher, to see the monkeys at his laboratory. "The children were very much amused," recorded Miss Littlefield, "and it gave a topic of conversation for some days following." At the end of the second year of the school, Miss Littlefield's mother died, and she had to resign to keep house for her father. Alec still had a dream that his school could work, but he needed to be inventive. With Miss Littlefield's responsibilities shared among the student-teachers who remained, he believed that the school could function for the interim.

In the summer of 1885, he and Mabel went on vacation with Melville and Eliza to visit Newfoundland, where Melville had spent time as a young man. On the way there, they stopped off at Cape Breton, Nova Scotia, where they reveled in the Bras d'Or Lake, an inland sea, and climbed hills that reminded Alec of his childhood romps up Corstorphine Hill. Alec and Mabel quickly fell in love with the landscape and the people.

But as they relished their escape, a new demand was crashing into Alec's life. That summer, two telephone companies, Pan-Electric and the National Improved Telephone Company of Louisiana, had joined forces in a joint petition for a government proceeding against Alec. While technically they were asking for a government suit, they were planning for it to be in name only. They expected they would still have control. At first, they alleged that his patent was generic and thus fraudulent because the telephone was not actually new, and thus they had the right to distribute their own telephones. Later they augmented this, alleging fraud in the patent office, an allegation supported by patent examiner Zenas Wilber, who began to speak on the record. By fall, the newspapers filled with coverage of the suit, and Alec was assailed by what he saw as attacks on his character. In early October, Alec and his family returned from Canada, and Alec immediately began to put together his counterattack.

On Monday, October 5, his school opened for its third year, but with no head teacher and Alec's attention elsewhere, the school was barely holding together. That Friday, the teachers administered an examination to the two most advanced deaf students. The whole school was based on the premise of teaching deaf children to function in the hearing world, and so Alec wanted to test what the students would do if someone unaccustomed to deaf people were to try to communicate with them. The teachers wrote several questions down on a piece of paper, and the children were directed, by speech and by gesture, to respond in the blank spaces that followed each question.

But what the students did was horrifying to Alec. Instead of responding to the questions asked, they simply copied what was written. His school was never meant to be theoretical; everything about deaf education, to him, was practical, about survival and self-empowerment. The idea of the questions was to mimic a real-life situation in which a student would need to interact with the hearing world without an intermediary. If they were lost, for example, a stranger would probably ask them the most basic questions: *What is your name? Where do you live?* In this situation, their test responses implied, they would have nothing to offer. Only the same words again. *What is your name? Where do you live?* It was beyond unacceptable to him. It was dangerous. And he was furious.

Alec did want the students at his school to learn to speak, but in practice, that was not his biggest goal. What he wanted was for the deaf to be able to function in society as well as hearing people. Speech, he was finally about to concede, was only a part of that.

His full articulation of that argument would come in response to several factors: this test; his students' broad lack of general knowledge and educational achievement; and finally, the insistence of the remaining teachers that speech was the highest goal of the school. One of his student teachers, Mrs. Bingham, saw need and want everywhere in these students' educations—but for her, and many of her peers, "nothing else is of so great importance" as speech.

She was, in many ways, a reflection of a larger force in oralism, one that wanted students to be able to culturally *become hearing*. Their voices should shake off all traces of a deaf accent. Their voices should be perfect.

She had embraced Alec's message of speech as a vehicle to social uplift. But he was the expert, and now Mrs. Bingham thought Alec's absence was to blame for the students' speech imperfections. And she wanted to know why he wasn't at the school.

It was the same story again: Alec wanted to be at the school, but there was the telephone, always the telephone. The petition for a government lawsuit had been taken up in Tennessee, before being swiftly shut down by President Grover Cleveland due to the amount of private interest involved by those overseeing the case; namely, Attorney General Augustus Garland was a shareholder in Pan-Electric. But the case wasn't over—it was merely being transformed. Instead of a case driven by private litigants, it was about to become a genuine Department of Justice suit. Alec's attentions were on his defense. And by now he felt he should be able to afford a brief absence from teaching. He had teachers, after all. Why they were languishing without him was a mystery to him.

What he wrote to Mrs. Bingham was a letter that could be repurposed just as easily, a century or more later, to argue against the very method he was trying to advance. He believed he was writing about only a few teachers, just the one school, but he underlines everything that would come to curse his system, and much of what would come to curse his name.

He wrote, "[The students] not only have mouths to be taught to speak and eyes to be taught to hear—but they have something of far higher importance—the cultivation of which has been cruelly and sadly neglected. They have minds to think, ideas to express—questions to ask. And have not we thoughts that we would give all we possess in the world to communicate to them? And we can not."

His words are both rectifying and damning. They mean that he saw with clarity what was important, that he had come to care about the minds of these deaf students, not only their voices. But it also means that he had the evidence before him, that he could see the way his method was going wrong—even in his own school, where he had the most power to wield. As always, he believed that the problem was not in the fundamental idea of oralism, but in the small elements he could not personally control. There were so many things he'd blamed for the failures of his past: individual students, the telephone, parents, the lack of commitment of other people, lack of time, lack of health. Now he believed his school was failing because of Mrs. Bingham, or other teachers, or the legal pressures on Bell Telephone, but the truth was that the school was built on oralism's weak foundation. Speech was not the most important thing. It was not then, and would never be, more important than the development of the mind.

At Milan, Susanna Hull had said: "I have found that with stronger faith in [speech], utter surrender of the mistaken desire for speedy knowledge, and more patient drill in the first elements of sound, failure cannot come." But she was dead wrong. Deaf people and their allies knew this both experientially and instinctively; if the deaf weren't taught to think, then speech would never matter. The mind had to come first. They didn't quite have the terminology for *why*—they didn't have the phrase *language deprivation* or an understanding of the science behind it—but there was a keen sense that under oralism the mind was under threat.

Even Mabel, a poster child of oralism, owed her success in part to her teacher's insistence on developing the mind. For Miss True, perfection of sounds was secondary to the reception of information. Mabel always learned at her grade level. In neglecting the mind, in making it secondary to speech, oralism irreversibly harmed its deaf students. This was what it was doing in Alec's school; it is what it would do to countless students for decades and decades to come.

Alec recognized the very core of what had become, and would continue to become, the greatest problem of oralism: the deaf children he wanted so desperately to liberate were in fact being damaged by his method. It may have been true that he was able to teach a handful of individual deaf students, in one-on-one lessons, how to speak. But such success had not proven replicable en masse, for the majority of students, by the majority of teachers.

Around this time, Gallaudet suggested that Alec, and those like him, were arguing "from the particular to the general." Alec had witnessed and worked with several students who could do the things he asked of them, perhaps most famously his own wife. But this did not mean that all students could do the same. It didn't even mean that his own students could always do so: George Sanders, Alec's first student in America, who had Alec's attentions at home—his doting care and his love—never did learn to speak particularly well. Now he was enrolled in the National Deaf-Mute College, the center of deaf intelligentsia.

At 2:00 p.m. on Wednesday, November 18, 1885, Alec gave a speech announcing the closure of his school at the end of the academic year. He was distressed by the disappointing results from even his most advanced students and didn't have the time or energy to train the teachers better or take on a greater share of the teaching. Mabel felt partially responsible, felt she should have done more to support his work. "His little school was the one great object of his life," she wrote.

It was a new low for Alec, a moment when he could have examined the failure of his school, could have come to see it as a failure of his method. He could have asked questions, could have wondered. But he didn't. Instead, he found something new, some*one* new, who would give him reason to hope. She was a young girl, deaf and blind, and already her parents were going through the same desperate motions that so often drove families to Alec's doorstep.

In this moment, Alec knew nothing of her, nor of the hope they would bring to each other. For now he could do nothing but give himself over to the disappointment of shuttering his beloved school. After his closing speech, he returned home and lay on the couch with "a splitting headache and a heartache harder to bear." He told Mabel that his whole life had been shipwrecked, that Mabel was all he had left.

Chapter 16

There is a pleasure in the pathless woods,
There is a rapture on the lonely shore,
There is society, where none intrudes,
By the deep sea, and music in its roar:
I love not man the less, but Nature more . . .

—Lord Byron, from "There Is a Pleasure in the Pathless Woods,"
a favorite poem of Alexander Graham Bell's

Three years earlier, in Tuscumbia, Alabama, in a two-room cottage covered in vines and honeysuckle, Helen Keller was expected to die. It was January 1882, only months after the death of President Garfield, and she was nineteen months old. Her family gathered around her bed and watched and waited. She'd been diagnosed with "brain fever"—either scarlet fever or meningitis. No one had much hope.

But then her health turned. After several days, she seemed stronger, more alert, though it only made her more aware of the things that were different. Her eyes were hot and dry; she turned them toward the wall and away from the light. What light she allowed herself to see grew dimmer and dimmer until one day she woke and all was dark. She awoke, she believed, into the night. She waited for the sun. She waited for light, and wondered why it took so long to come.

By the time her parents understood what had happened, Helen could no longer see, and she could no longer hear.

After announcing the closure of his school in late November 1885, Alec left his home in Washington, DC. He left his school, his pet monkeys, his lab, his wife and children. He wanted to leave everything behind. He went to Martha's Vineyard to return to his genealogical studies, investigating any deaf family in New England with two or more deaf children. He knew that the success he sought lay at the other side of hard, focused work.

He traveled simply as A. G. Bell, hoping to be incognito. His letters to Mabel stank of cigars, which she gently chastised him for. The island moved through the modest rhythms of a hard winter, quieted from the summer tourist hum. Alec stayed in "dull and cheerless" Cottage City, with miles of cottages abandoned for the season, "and the stillness of a Scotch Sabbath resting over all." In nearby Vineyard Haven, Alec went on hours-long walks, which he took in hopes of strengthening his health before his upcoming doctor's appointment.

He spent the rest of his time copying down genealogy records. Many in New England's deaf community had connections to Vineyard families; in the nineteenth century, approximately one in every 5,728 people were born deaf in America, but on Martha's Vineyard, the ratio was one in every 155. In the town of Chilmark, one in four residents was deaf. The island, which had the highest concentration of deafness in New England, was a gold mine for information on genetically deaf births, though the record books themselves were slowly disappearing. The books that remained were only the few that survived a fire, and their condition was bleak—faded, the corners rubbed out by thumbs of old clerks turning pages. "No wonder the leaves are ragged and dirty," wrote Alec. "No wonder the records are fading away."

As he was copying these record books, though, Alec was ignoring another phenomenon: in several towns on Martha's Vineyard, deafness wasn't a calamity but an afterthought, an easily accommodated piece of human diversity. The deaf married, had careers, went to church, had wide circles of friends, had families, and none of this was confined to a separate community; they were fully integrated into the hearing world. Or, more accurately, the deaf had integrated hearing people into their world. On the island, the hearing people signed. It was a utopia of sorts for deafness, the kind of utopia that Alec sought—deaf people and hearing people lived peacefully side by side. Whether someone was deaf or not was an afterthought in the memories of islanders. A deaf person was first renowned as "a great fisherman," remembered to be deaf only after some prodding into memory. One hearing woman, who was born into the community only shortly after Alec's visit, remembered that "those people weren't handicapped. They were just deaf."

But while Alec believed in hearing-deaf integration, his vision did not include the idea of hearing people learning sign language. He observed this phenomenon, but he didn't dwell on it. It didn't seem like a solution worth exploring.

Instead, he did what the telephone had taught him to do: push everything aside, work tirelessly and singularly, focus on the larger picture, the utopia he believed he could usher into being. He let the hum of his failure die down to a nearly inaudible din; he let himself forget.

Even as Alec was on the island to observe, record, and study genetic deafness, he was already set in his ideas about how to achieve deaf equality. He held fast to his promotion of oralism, even as he had begun to see the ways in which the broader education of these children was being sacrificed to teach them spoken English. He was back in touch with Sarah Fuller, and together they were working toward replicating his style of school in Boston, all while keeping their eyes open for students who could represent the success of the method. He forced himself to believe that the failure of oralism under his watch was only an isolated incident, a strange misstep in an experimental school. He pushed past it.

Within four days he'd copied 206 pages of records. He was supposed to leave Massachusetts and travel to New York by December 7, but these records, he felt, were too important to abandon. Besides, there were more records in Edgartown and deaf families he wanted to visit. "I had to decide—either to go and leave my work half done or stay and *finish it.*" He decided to stay.

He worked compulsively. For months he had been tracking the family trees of any deaf person close to him, and now he fleshed out the trees of deaf families whose names he'd been seeing for years. He copied and copied, collected data, and preserved it for future analysis. When he wasn't copying records, he spent two to three hours a day walking in the deep oak woods that surrounded the towns. They smelled of ocean air and crisp decay, and there was something mournful about the leaves fallen from the trees, something sad that he was drawn to, and yet something alive, too, in the bracing winter air. Alec took up quiet residence in the "Sailors' Free Reading Room" to copy the Mayhew family tree, which the Unitarian minister, Mr. Stevens, let him borrow. As Alec copied the records, he listened as the sailors told Mr. Stevens of the countries they traveled from, the cargo they were carrying; Mr. Stevens would help them find something to read—free copies of *Harper's Monthly*, *Scribner's Magazine*, assorted novels, and what Alec determined were "not a few religious tracts."

From outside the Sailors' Free Reading Room, Alec could hear the breaking of huge waves against the shore, though the water everywhere was calm as far as he could see. Mr. Stevens told him it was the waves of a storm, huge billows, not against this shore but the south shore of the island, caused by yesterday's storm. Astonished and enchanted, Alec stood there, on a graying old porch cluttered with whalebones and the figureheads of old whaling ships and listened to the breaking of yesterday's waves more than nine miles away.

He thought of Mabel, a memory from a decade before, when he first knew her: walnut furniture, warm summer air, and Mabel, seated at her mother's feet, Mabel's arm on her mother's knee, her head on her mother's lap. At the time of this memory, he had just learned that Mabel would be leaving for the summer, for Nantucket, not far from where he was now. He remembered her pale arms and her mother playing with her long hair.

And he remembered, too, how he hoped it would be if she could love him, how he would protect her from her deafness. "My great strong love would shield you," he later wrote to her, "and compass you round so that you should never know or realize the affliction that had clouded your life."

Looking back, it would seem that everything he'd tried to do since then had gone awry. He still believed that oralism wasn't respected enough, and the combinists still held significant sway. Combinism may not have had power in the wider hearing world, but it was still the dominant voice in the field. Besides, oralists didn't want combinists in the field at all—they were fighting for total dominance. And Alec believed that this dominance ought to have come years before now. Instead, his school was dissolved. His memoir was hated. He still wanted to eliminate discrimination against the deaf, but his mind increasingly went toward eliminating deafness itself. He wanted to do something big. He wanted, maybe more than anything, a legacy in the world of deafness, a single contribution that wouldn't fall apart.

In Alabama, Helen had found ways to communicate certain words—words like *yes* and *no*, *water* and *bread*—but she knew that she was missing out on something more. Other people moved their mouths for reasons she couldn't understand. She tried both: moving her lips and gesticulating as she had always done. Nothing worked. Her attempts at communication ended, increasingly, with kicking and screaming.

This would become part of the fundamental story of Helen: her childhood temper, her anger, what was so often described as her animalistic or monstrous nature. No one would question this, wonder what it would be like to have no way to control or make sense of the world. It wasn't the state of being languageless. It was animal. It was monster.

In the summer of 1886, when Helen was about six years old, she and her family made the journey to see a Baltimore doctor, whom they hoped might be able to do something for her eyes. On the train, Helen made friends: one lady gave her a box of shells, and Mr. Keller poked holes in them so Helen could string them together. The conductor let her hang on to his coattails while he went down the aisle punching tickets, and then gave Helen a punch to play with on her own. She sat in the corner of her seat for hours, punching holes in cardboard as the train made its way to Baltimore.

Not long after Alec returned from Martha's Vineyard, Secretary of the Interior L. Q. L. Lamar called for a genuine government suit against the Bell Telephone Company—one that would be controlled by the Department of Justice. This would cut the Pan-Electric and National Improved Telephone companies out of the suit altogether, thus eliminating their ability to settle, making for a purer inquiry into the validity of the Bell patents and determining once and for all the charges of patent infringement and fraud that were coming up against the company.

In addition to the government case, Pan-Electric had also elected to move forward with their suit on patent infringement, likely with the hopes of reaching a settlement. And another inventor, a poor Italian immigrant, Antonio Meucci, claimed that Alec had stolen his idea—he had filed a caveat for a telephone back in 1871. Storrow wrote to Alec not to worry about Meucci, whom he called "the silliest and weakest imposter who has ever turned up against the patent."

Meanwhile, several older court cases had now consolidated and made their way to the Supreme Court, which would hand down a definitive ruling on the company's rights to the invention. Arguments would be held in January and February of 1887. If they lost, the Bell Telephone Company stood to lose their patent, and with it, their monopoly.

Back in DC, as Alec prepared for court, his life returned to its usual rhythms. Families of deaf children came to find him, sometimes on pilgrimages from far-flung places, looking for the man they believed could offer salvation. They were often in the midst of a very specific grief, one of parents coming to terms with the ways their child's ability would be fundamentally different from what they'd expected. They were working to ensure that this didn't disrupt everything that they hoped their child could be and everything the child would someday want to be. They didn't see this as an issue of identity; they were desperate to stay close to their children and to protect them as best they knew how. They were adjusting their dreams, recalibrating their understandings of what was possible, and vulnerable to persuasion that they might be able to hang on to some semblance of who they thought their child would be.

Alec received correspondence from these families. Hearing relatives detailed the ages of the deaf child, how old she was when she lost her hearing, how quick she was to learn. They asked how much lessons would cost, how long they would take, if he would teach them. They worried their child would never be able to communicate again. Or, they asked for him to send advice—what should they do if they live in Africa, if theirs is the only deaf child in the area? The youngest deaf writers detailed what they liked to study: "I study the History, Geography, Arithmetic and Literature. And Metal and Minerals." Their parents hoped he would take their children on.

He heard, too, from former students, who wrote to thank him for paying for their classes ("I think it may help me to get work") or for furnishing a printing press for a deaf school ("I think it is very kind of you") or who wrote just to say that they watched the stock market in hopes that the telephone stocks were rising.

They wrote, and they came. They asked for education, jobs, money. Alec asked the National Bell Telephone Company to set aside some jobs specifically for deaf people, but they didn't. He often hired deaf people as printers or assistants, though never as teachers. He gave money to support their small businesses, and consistently paid for their educations and even for their school clothes. When one deaf man died, leaving his two young children penniless, Alec gave $15,000 to help support them.

Another time, when a deaf woman's wealthy father was trying to annul her marriage by having her declared an "imbecile," Alec wrote publicly in defense of the woman, arguing that ASL was as distinct a language as French or German, and that to rely on communication in a second language, English, to legally deduce her intelligence was wrong. "Justice demands that she should have the assistance of an interpreter, skilled in the use of the sign language," he wrote. These arguments—that ASL was a distinct and full language, that the deaf were entitled to interpretation—were a hundred years from even beginning to be taken as true by the general population. But while Alec believed that ASL was a language, and that adults were entitled to its use, he still didn't believe that it was helping anyone. He didn't believe it should be taught.

In Baltimore, the doctor, Julian Chisholm, could offer no help with Helen's blindness, but he did believe she could be helped, and he told her father to seek Alec's help. This was how they ended up in Washington, DC, on Alec's doorstep.

Though he didn't know ahead of time that they were coming, Alec invited them in, pulled young Helen onto his lap, and let her play with his watch. He understood her when she mimed a shiver, like the sign for cold. He knew she was referring to the icebox. She loved him immediately. Later, she wrote that the meeting was "the door through which I [passed] from darkness into light, from isolation to friendship, companionship, knowledge, love."

Alec invited Helen and her father to dinner to discuss her education. He sent Gallaudet a quick letter, saying that Helen was "evidently an intelligent child—and altogether this is such an interesting case—that I thought you would like to know about it," but Gallaudet was unable to meet with them. In the end, Alec advised Mr. Keller on how to arrange for Helen's education. He suggested he contact the then president of the Perkins Institution for the Blind, Michael Anagnos. Anagnos suggested enlisting the help of a woman who had gone to the school, Anne Sullivan.

Months later, back in Tuscumbia, Helen felt as though she were in a fog, inside of something both white and dark, both airy and tangible, through which she moved with no compass, no direction.

And then, one day, she felt from the bustle of the home that something was about to happen. She waited on the front steps, felt the sun shine through the honeysuckle, and played with the blossoms. She felt footsteps. A hand reached out, but it wasn't her mother. It pulled her in and held her close. It was March 3, 1887, and Anne Sullivan had arrived. She took her place at Helen's side, where she would stay, with only small interruptions, until her death.

On this first day, she brought with her a doll, a gift from the students at Perkins, dressed by Laura Bridgman, the famous DeafBlind girl taught by Howe. Anne lived at Perkins in the same cottage as Laura. Laura was in her late fifties by then, still living at the school, tucked away like a dirty secret, a miracle unfulfilled. Before Anne left for Alabama, Laura advised her on Helen's education and taught Anne the manual alphabet.

Now it would be Helen who would carry on the legacy of the educated DeafBlind girl, Helen whose accomplishments would eclipse all the poster children who came before. But in the beginning, Helen was none of this, and in the weeks that followed, Anne tried to teach her the simple notion that things had words.

m-u-g
p-i-n
h-a-t
s-t-a-n-d
w-a-t-e-r

Helen understood none of it, slid back into her temper. How could she tell the difference between the word for *water* and the word for *mug* when they were both, it seemed, the same? When everything was a nondescript flutter of fingers?

When the word was *d-o-l-l*, Helen threw the doll from Perkins to the floor and felt it break at her feet. Helen felt nothing but satisfaction as Anne swept it up and put the pieces to the side of the hearth. She had simply rid herself of the thing that bothered her.

And then, as that anger subsided, she and Anne followed the saccharine smell of honeysuckle to the well house, where Anne held Helen's hand under a steady stream of water. Into Helen's other hand, she spelled, slowly, *w-a-t-e-r*, *w-a-t-e-r*, *w-a-t-e-r*, and then faster, faster, one of Helen's hands with water flowing over it, the other, Anne's fingers: *water water waterwaterwater.*

Helen stood still, her body alert, her attention on her two hands, water, letters. And then her face began to change. She felt something return to her. She understood the letters to be connected to the substance, *w-a-t-e-r* to water.

And then, the words for everything else. "Everything had a name, and each name gave birth to a new thought. As we returned to the house every object which I touched seemed to quiver with life." And then, suddenly filled with remorse and regret, she picked up the pieces of her doll. She tried to reassemble it.

In November 1887, eight months after she learned the word for *water*, Alec received his first letter from Helen Keller, written in her own blocky handwriting.

Dear Mr. Bell,

I am glad to write you a letter. Father will send you a picture. I and father and aunt did go to see you in Washington. I did play with your watch. I do love you. I saw doctor in Washington. He looked at my eyes.

I can read stories in my book. I can write and spell and count. Good girl. My sister can walk and run. We do have fun with Jumbo. Prince is not good dog. He can not get birds. Rat did kill baby pigeons. I am sorry. Rat does not know wrong. I and mother and teacher will go to Boston in June. I will see little blind girls. Nancy will go with me. She is a good doll. Father will buy me lovely new watch. Cousin Anna gave me a pretty doll. Her name is Allie.

Good-by,
Helen Keller

Chapter 17

Alec talked genealogy all the time. He thinks that in the course of a hundred years, material will be gathered through Genealogical Societies from which important deductions can be made affecting the human race. . . . I am afraid I am not particularly interested in investigations that can only be used a hundred years hence.

—Mabel Hubbard Bell

When Alec thought of what it meant to be deaf, he thought of loneliness. And not a walk alone in country fields; he understood, too, the difference between solitude and loneliness. He believed that the worst loneliness was that of being one among the many. He could picture it perfectly: a moment from his boyhood, standing on a sidewalk in the heart of London. Hundreds, thousands of faces passed by, and he didn't know a single soul. This was how he imagined the loneliness of the deaf. Watching hundreds of faces pass by every day, and not being able to communicate with a single one.

To save the deaf from this imagined fate, Alec spent increasing amounts of time alone, researching the history of individual deaf families and the inheritance of their deafness, trying to flesh out his theory of the ever-expanding deaf race. He seemed to think nothing of his own aloneness, not any more than he acknowledged the way signing deaf people found great community in each other, how they had organized schools, newspapers, a national advocacy organization. Even Mabel, the woman he most looked to as an example of self-sufficiency, wrote more faithfully to his mother, Eliza, than almost anyone else.

But then, Mabel wasn't able to access the broader deaf community. For the most part she didn't want to. When Alec befriended Albert Ballin randomly in a Paris bar, long before Ballin would write his anti-oralist *The Deaf Mute Howls*, the two communicated through finger spelling and signing, forming a fast friendship. But Ballin couldn't connect so easily with Mabel, who could neither finger spell nor sign. "In conversing with her," he wrote, "I moved my mouth without making any sound, and she always answered in writing. . . . She understood every word I spoke, if I moved my mouth broadly, deliberately, and if the topics were commonplace." But she struggled with harder words, or foreign words. When they traveled, Mabel relied on Daisy to interpret the captain's orders or what was said at dinner tables. Mabel's closest community was limited to those who could understand her speech. At home, this was primarily her family, and her family was primarily Alec, who was gone more and more of the time.

Mabel felt that she rarely had Alec's attention, and she missed him. "I wonder," she wrote, "do you ever think of me in the midst of that work of yours which I am so proud and yet so jealous, for I know it has stolen from me a part of my husband's heart, for where his thoughts and interests lie, there too must his heart be."

Mabel increasingly questioned his work—not just the way it took him away from her, but the way it revealed something about him to her. When he was running his school, she had pivoted her energies to help him run it, especially when he was away. His school, she could see, was connected to individual students, and so on some level she could support it. But this new work was different, bigger, darker. His work was now about ensuring that fewer deaf people were part of humanity.

As Alec became more powerful, he saw individual deaf people primarily as people who needed help, and his interactions with them were becoming more limited to this "helping." He worked less with deaf students, spent less time sharing a drink with deaf people he came across in bars. Deaf people reached out to him, but more often their letters went unanswered. It was Mabel who tethered him to the ground, who connected him to others. "You are always so thoughtful of others," he wrote, "whereas I somehow or other appear to be more interested in things than people—in people wholesale, rather than in person individual." Alec may have believed that the deaf needed a hearing person to wed them to the hearing world—but it was a deaf woman, Mabel, who did this for him. "I feel more and more as I grow older the tendency to retire into myself and be alone with my thoughts . . . ," wrote Alec to Mabel, "and you my dear constitute the chief link between myself and the world outside." When he was absorbed in work, it was she who insisted he go out and meet with friends and neighbors. "My deaf-mute researches have taken me away—far away—from you all," wrote Alec. "I don't think your thoughts—or feel your feelings—nothing but deaf-mute—deaf-mute—and solitude in my mind."

Through the mid- to late 1880s, Alec worked on his genetic studies, and traveled the country to promote the idea of day schools for the deaf. In 1888, he collaborated once again with Thomas Sanders and Gardiner Greene Hubbard to begin a new society for the exchange of ideas on science and culture. They called it the National Geographic Society. Gardiner became the society's first president, serving for close to a decade, though under his leadership it was a staid academic society and publication. When the presidency passed to Alec, in 1897, he reinvented the magazine for the general public, full of photographs, storytelling, and nontechnical language.

He also became increasingly entangled in both the Supreme Court case and the Pan-Electric case. By then, patent examiner Zenas Wilber had provided multiple affidavits admitting that he had shown Elisha Gray's caveat to Alec, but how well they would hold up was under question. Deeply entrenched in his alcoholism, Wilber had a tendency to give testimonies that contradicted each other. Regardless, Alec was preparing for a real fight to defend both his name and the company that bore it. He collected his papers from the time of his life when he was inventing, and assembled a defense. In 1888, the Supreme Court found the infringement cases for Alec, stating that he met the demands of describing the invention and what it would be used for, and it wasn't necessary for the invention to be in perfect operation in order to be valid. Even though this finding robbed the Pan-Electric case of their greatest arguments, the case continued to drag slowly on, collecting testimony until 1895. It ended, without judicial decision, with the 1896 death of the last government attorney left on the case.

Over a century later, in 2002, the US Congress would pass a resolution that officially acknowledged Meucci—the Italian immigrant Alec's lawyer called the "silliest and weakest imposter"—for conceptually predating Alec in the idea of the telephone. But that was years after they were all dead, years after Bell's name was solidified in history books.

In his own life, Alec's legal worries ceased after the Supreme Court case was settled. He turned to filling pages upon pages with elaborate family trees of families with deafness.

As Alec was pursuing this question of heredity in his own research, he was also continuing his broader public campaign to discourage deaf intermarriage. He presented his study at the National Academy of Sciences in 1885, 1886, and 1887, and *Science* magazine published prolifically on the subject.

At the 1889 convention of the National Association for the Deaf, then president E. A. Hodgson delivered an opening address that cited the gathering of statistics on deaf marriage—and the public concern they provoked—as one of the most important issues facing the convention. "Incidentally," he said, "this information may be used to either disprove or verify what, to most of us, seems an absurd theory on the danger of deaf-mute intermarriages. We must settle forever the sensational alarm concerning 'the formation of a deaf variety of the human race,' not with assertions only, but by an array of evidence that will cause Prof. Bell to haul down the danger-signal he has hoisted, and free us from the incubus of what is becoming a widespread public prejudice."

He suggested they collect data on the methods used to educate deaf individuals, to see if there was, as Alec proposed, any correlation between the educational method used and the propensity of a deaf person to marry another deaf person as well as, more broadly, the extent of their social interactions with hearing people.

Edward Allen Fay, the editor of *The American Annals of the Deaf* and vice president of the National Deaf-Mute College, read Hodgson's paper orally as Hodgson signed it, and Hodgson's suggestions must have taken hold. In 1889, Fay began his own extension of Alec's study, looking at the marriages of the deaf and the correlations of different types of deaf education. As soon as Alec heard—and with no apparent concern for Fay's support of combinist educations—Alec gave Fay access to all of his research, provided him with a workspace in the Volta Lab, and gave him the financial means by which to expand on the study as he saw fit.

Then, in the spring of 1890, the question pivoted into Alec's personal sphere. Mrs. Sanders, little George Sanders's grandmother, died. Mrs. Sanders, who had mended his clothes when they wore through, who had thought he had brain fever when really he was in love, who had cut his candles short when he would work too late into the night.

He had been back in touch with her recently. His work to keep George away from the deaf community was failing, and George had fallen in love with a deaf woman, Lucy Swett. Alec believed that when two deaf people intended to marry, friends and family should exert pressure on them to prevent it, and so he had contacted Mrs. Sanders, who begged George to reconsider. She died before she could change his mind.

On the day of the services, May 5, Alec made his way back to that white clapboard house with green shutters at the same time as Lucy was making her way away from it. They agreed to meet afterward.

Inside, Mrs. Sanders's body was laid out in the same room Alec had once called his study. Alec, at age forty-three, felt suddenly old.

The coffin itself rested in the same place Alec had once kissed Mabel, early on, with tears in his eyes. Back then his love had seemed hopeless, and in its own ways, it felt hopeless again. Mabel, at thirty-two, still looked startlingly young. Her face was still narrow, her skin taut, her hair glossy, her waist still cinched by a corset. As for Alec, gray had started to appear in two broad stripes down his beard and shimmered through his hair. He was growing rounder, and more tired.

"I remain solitary and alone," he wrote to Mabel later that day, "and every year takes me further from you and my friends and the world—and I seem powerless to help. . . . I hang like a dead weight on your young life."

After the services, Alec drove Lucy back to her family's home in Beverly and stayed for dinner. Though Lucy had been educated among hearing people, in hearing schools, she'd also grown up in a signing household, in a family of intergenerational deafness. At home, both of her parents were signers.

Back when Mrs. Sanders first said George and Lucy shouldn't marry, Lucy was thrown into despair. She asked George to tell her everything his grandmother had told him and then sent him away. She stopped eating, stopped sleeping. She believed it was true—she could not marry him. It would break her heart, she wrote, to have a deaf child. But it was breaking her heart more to imagine her life without George.

Lucy did the calculations. She went over it in her head, some impossible equation: her deaf great-grandfather had married a hearing woman and they had two deaf children, one of them Lucy's grandmother. Lucy's grandmother had married a hearing man and they had two hearing sons and one deaf son. Then one of the hearing sons went deaf at age eleven. He married a deaf woman. These were Lucy's parents. They had five hearing children. Two of them, including Lucy, became deaf.

But there was no math that could provide an answer, and eventually, Lucy could take it no longer. She cried from her longing to see George, to tell him she loved him. She had come to a realization: "George would marry a deaf lady anyway; Why should it not be me?" Even so, her reversal wasn't simple or easy. "We ought never marry, I know," she wrote. She was far from convinced that it was the right decision, but she knew that it was the only one she could make. She had given over to it, to her happiness and her love.

Now Alec and Lucy had a chance to talk directly, but Lucy was still unmovable. Alec became convinced that she truly loved George, and he could see why George loved her. When George kissed her in front of everyone, Alec knew he could hold no sway. Their love was more powerful than his statistics.

"They will surely marry," he wrote, "but what then? Will lovers ever consider the good of those that will come after them? Deafness has come down through four generations to Miss Swett, yet prudence will not prevent her from marrying one who is deaf—and George chooses danger to his offspring—for her love. Yet I can understand it too."

Maybe Alec could understand George and Lucy's choice to go forward with their marriage, but he couldn't let it go. On March 6, 1891, Alec was invited to give a talk at the National Deaf-Mute College. Knowing that George would be present, Alec had chosen his topic accordingly. He would speak to the students on marriage, just as he had spoken to George and Lucy individually, as he had spoken to the National Academy of Sciences, as he had spoken to the whole of the deaf world.

Standing before the gathering of students at the college, he promised the students he did not advocate for legislative interference—no matter what the rumors were: "I have no intention of interfering with your liberty of marriage. You can marry whom you choose, and I hope you will be happy. It is not for me to blame you for marrying to suit yourselves, for you all know that I myself, son of a deaf mother, have married a deaf wife."

In that room filled with members of the college's literary society, the students must have held their breath. Alec's proposal regarding marriage was already widely known in the community, and it was almost uniformly poorly received. A few years earlier, a man named Hiram Phelps Arms wrote in response to Alec, "As to the offspring being deaf and mute like their parents, 'what of that?' . . . Being themselves happy, well educated, finding in the great world around them plenty of occupations to engage their minds and hearts, one may well ask, 'What matters it' if they are deaf and mute. The orbit they may move in may have less grandeur in its sweep than that of those who hear and who are in the same social scale. But does it imply that the orbit of a deaf-mute must always define a lesser space? Assuredly not!" Arms represented the voice of a growing deaf intelligentsia, one that was educated, bilingual, and had a sense of power and identity. More than that, they were a group turning firmly against Alec. And here, they were gathered before him.

Alec continued, "I think, however, that it is the duty of every good man and every good woman to remember that children follow marriage, and I am sure that there is no one among the deaf who desires to have his affliction handed down to his children."

He wanted to empower deaf people with the knowledge of how to prevent more of themselves. He assumed the deaf also wanted this, that his logic was sturdy, and that these deaf students gathered before him—whom he called "the brightest and most intelligent minds among the deaf"—would help him spread the word. It was the note on which he ended his speech. He saw no inherent tension between his message and its audience. He saw nothing worth questioning in his mission: to rescue the deaf from loneliness by dismantling the strongest community they had.

That community meant nothing to Alec. By this point in his life, his primary connection to the deaf community was through the hearing people who occupied the places of stature in conjunction with it. While those individuals sometimes, though not always, engaged with the deaf community, Alec himself rarely did. If he did, it was most often from a place of paternalistic remove: teacher, benefactor, lecturer. In 1898, he became a trustee of the Clarke School for the Deaf. His stature grew, as did his distance from deaf people. Less and less often was he ever a minority in the midst of this community that was gaining their own political power.

Deaf people, however, were increasingly galvanized by Alec's words. They spoke up against him, arguing for their right to marry whom they choose. They were always careful to acknowledge the happy marriages that existed between hearing and deaf people but insisted that this happiness was also possible—even more likely—if both were deaf. D. W. George, NAD's third president, wrote, "Hearing people marry whom they love best and the deaf are entitled to the same privilege. If the deaf love the deaf best, why [not] [*sic*] let them marry and be happy. The affection which the deaf almost universally have for one another is undoubtedly the work of God."

It wasn't just amid the deaf community that he was surrounded by resistance to his most fundamental ideas—it was within his own life, within his own family. It was true that, though Mabel herself lost her hearing due to fever, deafness ran in her family. It was also true, though unsuspected, that Eliza's recessive hearing loss may have been genetic. If Alec had known, could it have stopped him? Would it have? His parents hadn't stopped him; neither had hers. Neither distance nor illness nor deluge stopped him. Neither class nor position nor his own pennilessness. Would he have worried enough for his unborn children to forsake his love for Mabel? Every piece of evidence that comes from his intellectual endeavors says yes, but everything about their story resists it. His love for her, from the beginning, was overwhelming. It's hard to believe that this man—this man of the train and the ferries, of the sleepless nights and headaches, this man of the thirty-three-page love letter—it's hard to believe that this man would expect anyone else to control their love.

But he did expect it, and his distance from his own wife fed into it. Alec was right that Mabel constituted his primary link to the outside world. It wasn't just the way she dragged him to social outings; Mabel was also the voice of empathetic reason, a quiet prodding to understand not only the facts around him but also the people, their fullness, their complexity.

Now Mabel said almost nothing about any of Alec's work. Only that—as he stood behind podiums wearing his dark woolen suits and delivering lectures filled with warnings about a deaf race—she missed him. She was involved with the Washington elite, and attended a few meetings of the "Washington Club" made up of eminent women writers, politicians, and artists, lip-reading furiously to keep up. But she never felt wholly comfortable in Washington circles. When Alec was away, she wrote to him loyally, filling him in on the minutiae of her days and pleas for him to return: "I like you around," she wrote, "even if you are a bother sometimes. You are the mainspring of my life, and though when it goes the other wheels go on by themselves for a time, it is very languidly and more slowly, and I want you back to give me an interest in life."

Her life was now split between their homes in Washington, DC, and Beinn Bhreagh, their Baddeck, Nova Scotia, estate, though they increasingly spent more time in Canada. She had imported the Washington Club into the community where she felt more comfortable, and where the women were more isolated and in need of such a thing. She called it the Young Ladies Club. She loved the days when her family was all together in Canada, where she could have boys' clothes specially made to fit her daughters so that they could wear pants and climb trees; where her favorite gift was an ax; and where Alec worked not on deafness but on sheep breeding, boatbuilding, and kite flying. Mabel described him for Eliza as "a big burly figure in his familiar knicker-bockers, which make him look like a big school-boy among the Lilliputians." How different he was in Canada from the man who stood at podiums and gave speeches on deafness and marriage.

And how different were they both from how it was in the beginning, when he traveled to Nantucket to find her, to deliver her a letter. Back then his actions were tempered by doubt. He believed he was doing the right thing, or the rightest thing he could do, but he didn't believe he could control the outcome. He believed he had to know and respect what she thought, that their future would be written together.

She did not go down to meet him that day in Nantucket, but from a window she might have seen him: his black hair askew, his dark eyes shining. Everything about him naked for her, his desire and love, and underneath that, his hope that he could offer her something. He could offer companionship, but this, too: he wanted to make the world different for her.

She couldn't have imagined this was how he would do it.

He had wanted to save her from ever knowing her own deafness—*My great strong love would shield you*—but her deafness was a part of her. She may not have communicated through ASL, but her experience was not the same as someone who could hear. This was something that Alec did his best not to acknowledge. And it was something she worked her whole life to leave unacknowledged.

Alec put the onus of change on the deaf—they had to control their voices, their language, their love—but deaf people had a different kind of change in mind. There was another idea that had grown exponentially in strength since the dawn of oralism, one that placed itself largely in opposition to oralism and broader hearing-centric ideologies, and that was that the deaf were simply a distinct culture, with a common language and common values. They didn't need to be fixed; all they needed was a chance. In a slightly altered social environment, they wouldn't even have to be all that different. They could just be themselves, and powerful.

It had been over thirty years since the National Deaf-Mute College had opened, and now there were generations of college-educated deaf people who were more than prepared to wrest their fates back from a man who seemed only passingly interested in what they thought about his ideas.

George and Lucy married. George, like many deaf people educated orally, never did learn how to speak very well. George and Lucy went on to have two daughters, Dorothy and Margaret, both of whom lost their hearing early in life. George and Lucy chose to teach them orally, to never use ASL with them when they were young. But they didn't oppose their eventual exposure to ASL, either.

And though Alec had once seen George as the son he never had, the two grew more and more distant over the years. All of Alec's attempts to save George had failed, at least in the way that Alec imagined it. George's love, his community, his deafness—this is where George found his strength. Alec couldn't recognize it. He could only turn away.

Chapter 18

Your deaf mute business is hardly human to you. You are very tender and gentle to the deaf children, but their interest to you lies in their being deaf, not in their humanity . . .

—Mabel Hubbard Bell

Sarah Fuller held Helen's hand over her lips and throat and let her feel her tongue as it made the sounds of speech. Within an hour, Helen had learned *M*, *P*, *A*, *S*, *T*, and *I*. Her first spoken sentence was, "It is warm."

The manual alphabet had opened Helen's mind, but she still felt what she called a sense of narrowness. "My thoughts would often rise and beat up like birds against the wind." She wanted to speak. Anne believed that too much emphasis was placed on teaching articulation to deaf children, but then one of Laura Bridgman's former teachers told Helen about a Norwegian girl, Ragnhild Kaata, who was DeafBlind and who spoke. From that moment on, nothing would stop Helen. In 1890, when Helen was nine years old, Anne took her to Boston to begin lessons with Miss Fuller—the same teacher who had partnered with Alec when he first landed on American soil, who first learned how to teach speech from Alec's methods.

Two years earlier, in 1888, Helen had made her first visit to the Perkins Institution, where Anne took her to meet Laura, who was sitting by a window crocheting a bit of lace. When Anne held her hand out, Laura recognized her instantly, and then kissed Helen in greeting.

Helen was instantly enthralled with Laura, but Laura was strict with Helen, instructing her to keep her hands clean and not to sit on the floor. Laura, who was exceedingly neat and clean, shrank from Helen's hands on her face, "like a mimosa blossom from my peering fingers," wrote Helen later. Laura's hands, reported Helen, were "beautiful, finely formed, delicate, and expressive," and Laura found Helen's movements too strong.

"You have not taught her to be very gentle," she said to Anne.

Helen found her to be "like a statue I had once felt in a garden, she was so motionless, and her hands were so cool, like flowers that have grown in shady places."

As different as Helen was from Laura, they were also the same in ways almost no one could understand. Helen observed that Samuel Gridley Howe had described Laura as "a well-formed child with a nervous, sanguine temperament, a large and beautifully shaped head, healthy and active," and goes on to observe that Laura described her with almost the same words.

It wasn't just Laura, though. When Helen arrived at Perkins, she thrilled to learn that most of the other children also knew how to communicate with their hands. "What joy to talk with other children in my own language!" she wrote. "Until then I had been like a foreigner speaking through an interpreter." At Perkins, Helen had her first experience of being able to communicate with almost anyone around, in a community where people could understand her with ease. It was like nothing she'd ever experienced before. "In the school where Laura Bridgman was taught," Helen wrote, "I was in my own country."

But just as she found the people with whom she felt most immediately connected, she was also about to be singled out in a new way. In the year of this first visit, Perkins Institution president Michael Anagnos wrote about Helen in the school's annual report: "As soon as a slight crevice was opened in the outer wall of their twofold imprisonment, her mental faculties emerged full-armed from their living tomb." With this, Helen Keller began to take shape in the public imagination.

People were enthralled with Helen and eager to glom on to the idea of the rescue of a young DeafBlind girl, especially if they could pair her not with Laura, who was also DeafBlind, or Anne, who was blind, but with a kindhearted hearing and sighted man, like Anagnos or Alec. Anagnos must have seen in Helen the possibility to drum up support for Perkins. Soon, the part of the report that spoke of Helen was reprinted as a standalone pamphlet, as well as in *Lend A Hand* magazine. Alec soon followed suit, sending a letter from Helen and a photo of her to a New York newspaper and also publishing them in *Science,* which he owned. He did this without consulting the Kellers, which irritated them, though they knew they had no recourse. Publishing these letters, he succeeded in both spreading the word about Helen and securing in the public's eye a link between them.

In the years since, Helen's story had spread, and not just within the realm of educators of the deaf and the blind. Amid a flurry of media attention, people began to travel to Helen's Tuscumbia home to see her; when she traveled north, she drew crowds wherever she appeared; soon enough she had been invited to meet President Cleveland. The public loved a miracle, and no one seemed more miraculous than Helen Keller—and the kindhearted hearing and sighted men they believed to be helping her.

In 1890, after a handful of speech lessons with Miss Fuller, Helen was speaking to trees, stones, birds. She spoke to Miss Fuller and Anne, who could understand her words, though no one else could, not for a long time. She practiced night and day, and whenever she felt discouraged, which was often, she would imagine being home again and speaking with her family.

Alec had been following Helen's education since their first meeting, and now Miss Fuller wrote to him that Helen's voice was "sweet and natural," that her speaking ability was on par with any other deaf child. Later, after nine total lessons, Miss Fuller beamed that "her ability to speak is almost marvelous."

The next year, 1891, Alec distributed a slender book at the conference of his newfound American Association to Promote the Teaching of Speech to the Deaf, or AAPTSD for short. It told the story of Helen, blind and deaf, who learned to speak.

Now Alec had two organizations, the AAPTSD and the Volta Bureau, an outgrowth of his science-based Volta Lab. The Volta Lab held the patent for the graphophone, an improved phonograph, which made a small fortune. After this, Alec's involvement in the work of the Volta Lab was minimal. He took his portion of the graphophone profits, twenty times the amount of the original Volta prize, and invested it into a new organization, the Volta Bureau, which focused on research into the deaf.

The organizations were housed at the back of Melville's DC home until 1893, when he and Alec collaborated to fund the construction of the Volta Bureau building, across the street. While the Volta Bureau took care of research, the AAPTSD was more about promotion—and Helen Keller was a promotioner's dream.

The book was so popular that it went into a second printing, which included a note from Alec: "The more attention can be directed to the fact that a child—blind and deaf from infancy—has been taught to think in the English language, and to read and write and speak with fluency, the more will the public be prepared to realize that deaf children who are not blind may be similarly taught."

It was the perfect souvenir to spread the gospel of speech, and it would do for Alec's ideas what Laura Bridgman did for Samuel Gridley Howe's. Helen was part of his master strategy, and she spent more and more time with his family.

At the Bells', Mabel read the lips of all the visitors who came to the house—certainly, thought Helen, "she needed patience, skill, and humor" to achieve this. But she also notes that Mabel, who believed that sign language isolated the deaf from "normal people," never did learn even so much as finger spelling. She doesn't mention the way this isolated Mabel from her. Still, Helen remembers Mabel running her hands over pieces of delicate lace. She would "hold a filmy web in her hands and show me how to trace the woven flowers and leaves, the saucy Cupids, the silken winding streams, and the lacy criss-cross of fairy paths bordered by aërial boughs."

Mostly, Helen spent time with Elsie and Daisy, and especially Daisy, who was her age and who "tried to put all the bright things she heard into my hand so I could laugh with her."

Still, Mabel, with her carefully measured voice of conscience, brought up her hesitancy about the whole situation. "When you come to think of it," she wrote to her husband, "all this is very selfish, we have been working Helen for all she is worth for our benefit and not hers. I hope that she will get some good out of it all." She left it at that, but it was also true that she was always on the lookout for this: deaf teachers who saw deaf students as experiments more than humans.

She was right to be hesitant—it was unethical not only to Helen but also to deaf children more broadly. Those who "succeeded" at oralism were, like Helen, outliers. For every child who learned to speak, there were nine children who struggled, who were set back more and more with each passing month of their oralist educations. That didn't make for a good story, though. Helen Keller did. Her story proved the possibility of great success, the possibility of miracles. And it would make it all the more difficult for those nine out of ten, all of whom would be expected to become miracles of the same magnitude.

Meanwhile, Alec's efforts to wipe out sign language festered in the deaf community. It was not necessarily that the deaf didn't want the ability to speak or lip-read, but the extremism of the oralist approach was alarming. In 1895, Gallaudet professor Amos Draper summarized the perspective of his community: "[Deaf people] would in very many instances appear in support of [speech and lip-reading] were they not placed on the defensive as against claims of pure oralists which they think aggressive in spirit, idealistic rather than practical, and not supported by their results, viewed broadly and apart from special cases."

What they did believe in was the value of having as much access as possible: "They believe that a manually taught deaf man is all the better off if he has any modicum of speech. They believe that a pure-orally taught deaf man, even if he reads the lips like a prescient angel, is better off for knowing the manual alphabet. They believe that perhaps all the deaf, and certainly the vast majority of them, receive untold aid and comfort through the sign-language."

What deaf people preferred was to have the opportunity to learn both through speech and signs, depending on the abilities and desires of individual pupils. This system, the combined system, was fast displacing both pure oralist and pure manualist schools.

By 1895, manualist schools were nearly extinct—only four remained open, with only 199 students total. Most deaf schools, sixty-three of them, were combinist, and most students, 7,906, were in those schools. Oralist schools were trailing behind this movement, with only twenty schools and 1,265 students. But this didn't mean oralism wasn't doing harm. Including the students in combinist schools educated under solely oralist methods, the total number of orally educated children was 2,415, or 23 percent of the total number of deaf children in schools. Oralism was still threatening the deaf community—and now educators gathered together in Flint, Michigan, to discuss it.

Temperatures hovered in the nineties in July 1895, as deaf educators made their way to Flint on the train. They paid a small amount of money to stay at the Michigan School for the Deaf, where the annual Convention of American Instructors of the Deaf (CAID) was being held. It was a professional convention, but almost a hundred local deaf people were in attendance, and frustrations with Alec were especially high.

It hadn't always been that way, though. As recently as six years ago, at the third convention of the National Association of the Deaf, Gallaudet had stood up for Alec publicly. He encouraged the association to "cooperate with all the intelligent men who are laboring to help the deaf," and added that "Professor Bell is one of these. He is a noble man, who gives his time and money for the benefit of the deaf. . . . Prof. Bell is the friend of the deaf. Meet his theories by facts and prove them wrong. He is sincere and generous, full of enthusiasm for all that he does." But that was six years ago, and this was now.

By now the impacts of oralism on adults in the community were vivid. As oralism spread, deaf educators of the deaf were pushed out of their teaching positions; Amos Draper had expressed this concern about the method, too: "It asks all deaf persons now employed [as teachers] to leave, and, moreover, to yield their support to a plan which shall bar out all deaf persons who might in time succeed them. Is this not asking much of the deaf?" This was harmful on the level of these individual deaf teachers, but Draper conceded that they would possibly choose to make this sacrifice—but only if oralists could demonstrate that "pure oralism makes the great mass of the deaf wiser, happier, and more prosperous than any other method or combination of methods." He had not yet seen any such proof. And as adult deaf people were removed from these positions, deaf children were being denied regular access to deaf role models.

Leading up to the conference, Gallaudet had been trying to secure public funding to open a normal school at the National Deaf-Mute College, and Alec had been fighting him tooth and nail. Alec believed that public funds shouldn't support one method of education over another, and that whatever normal school would open at the college would have to support combinist educations, since he believed the deaf themselves couldn't be trained to teach speech.

Gallaudet promised Alec that he wouldn't admit deaf students to the normal school—problematic in and of itself, to be sure—but Alec didn't believe him. Instead, he saw Gallaudet's move as an affront to his methods and spoke before Congress to oppose the allocation of funds to the college. Alec lost that battle, and now tensions between the two men were especially high.

On Tuesday, July 2, at 2:00 p.m., the convention was called into session. There were addresses of welcome, general reports, and socializing well into the evening.

Meanwhile, Alec was still making the long trip from Beinn Bhreagh to Flint. As he did, Gardiner wrote to his son-in-law, anxious about how the convention would go: "I do not feel certain as to what Dr. Gallaudet will do + am sorry that I cannot be there to help if necessary."

Alec was worried, too. For him, the number of deaf people present—close to one hundred—was the sticking point. It was enough, he said, "to swamp the votes of all the [superintendents] and [principals] present." It was enough, in other words, to put the future of deaf education into the hands of the deaf.

Mabel, too, was worried, though not so much about the conference. She was in Paris, receiving her husband's letters at a two-week delay. From what she could tell, Alec was cutting himself off again; "the very nuns here are not leading as solitary narrow a life as you," she wrote. She was right. Even in the midst of a massive gathering, Alec kept his distance from everyone else. When he arrived, he went to the Bryant House, about a mile and a half away from the school.

His first morning there, he ate breakfast at 8:00 a.m., and was at the institution a little after 9:00 a.m. It was the day Gallaudet, voted president of the newly reorganized CAID, was scheduled to present a paper, its topic unnamed. The members crowded into the lecture hall. In the July heat, the temperature in that unventilated hall began to rise.

A blackboard ran the length of the stage, and Edward Miner Gallaudet and his ASL interpreter faced the wooden hall, full of teachers of the deaf. Edward began his speech, noting what appeared to be the progresses of the past five years, the harmony that had been built among all those in the profession, the sense of camaraderie and cooperation.

But the truth, as he saw it, was that there was a threat to deaf education, represented by a man with "peculiar views," a "partisan spirit," and "a persistency out of all reason."

No one could have predicted that "definite and strenuous efforts would be made by Prof. Bell . . . to push the cause of pure-oralism to the fore, and to secure for it, as soon as possible, a controlling influence in the work of deaf-mute education in America." In this, he said, Alec and other oralists "have worked with a partisan spirit and purpose, calculated to engender serious if not permanent antagonism in the profession." Frankly, Gallaudet had had enough of it. He was no longer interested in playing nice.

After dredging up all manner of personal and professional conflicts with Alec, Edward poised himself for the final blow, the final mockery. He started on one side of the stage to finger spell, one letter at a time, while walking the length of the stage, the cumbersome name of Alec's Association: *T-h-e A-m-e-r-i-c-a-n A-s-s-o-c-i-a-t-i-o-n f-o-r t-h-e P-r-o-m-o-t-i-o-n o-f S-p-e-e-c-h t-o t-h-e D-e-a-f.*

Members of the audience snickered, quietly at first, but then louder. Those who believed in Alec's methods, who taught with those methods, sat in stern disapproval as the heat of the hall crept close to a hundred degrees, if not warmer.

If Alec had any hopes of being taken seriously by the broader community of deaf educators, they were dashed. And though these were Gallaudet's words, they represented a larger growing sentiment. Even Mabel was worried about Alec—both his increasing seclusion and the way it might be skewing his work.

Around this time, she wrote to him: "I want you to succeed in your experiments, but not to lose all human interest in the process. Your deaf mute business is hardly human to you. You are very tender and gentle to the deaf children, but their interest to you lies in their being deaf not in their humanity . . ."

She was noting the quality that would become Alec's greatest downfall: his detachment from the human element of the work, ironically seated in his tendency to set himself apart, the very thing that he wanted to save the deaf from.

But here at this conference, the deaf were not isolated. The conference itself was just one of many demonstrations of the strength of their community, full of deaf intellectuals and educators. But Alec found his community apart from the main event, at his hotel. After the rest of the day's events and dinner, he returned to his hotel with a group of principals and teachers. They stayed up smoking and fuming until around 2:00 a.m. "Heat tremendous," wrote Alec, "external and *internal*."

Officially, the CAID was to embrace a variety of methods. In the *American Annals of the Deaf*, Gallaudet wrote that their platform "urges that every deaf child should have an opportunity, under the most favorable possible conditions, to learn to speak; . . . at the same time acknowledging the pregnant fact that a variety of capability exists among the deaf which *compels* variety in method." The combinists were giving up entirely on the idea of further compromise.

The blow of the speech drove an intractable wedge between the two men, and between the two schools of thought. Now Alec was truly on his own. But the fight for dominance was not over, and Alec wasn't done fighting.

The next year, in Mt. Airy, Philadelphia, Helen took to the stage.

By then, Helen was fifteen. She wore her hair coiled into ringlets, and the corners of her mouth often lifted into a soft smile as she listened to another's descriptions of the world with either her hands on another's to read their finger spelling, or else with her fingers on their face to read their lips. Anne was still her constant companion, and Alec buoyed them both—later in life, when Anne was asked how she went through so many years by Helen's side in tireless labor, she replied, "I think it must have been Dr. Bell—his faith in me." Alec had invited Anne to present a paper at the conference, but the last time she tried to present to the AAPTSD, she panicked at the last minute and Alec read the paper for her. This time, she turned down the offer. Instead, all attention was on Helen.

There were about four hundred people gathered that day, and Helen stood before them all. She had prepared a paper titled "The Value of Speech to the Deaf," and used her voice—trained by Sarah Fuller, who was trained by Alec—to impress upon her audience what a difference speech had made in her life. She knew there was great disagreement about whether the deaf should learn to speak, but frankly, she couldn't figure out why. She loved speaking. "It brings me into [a] closer and tenderer relationship with those I love, and makes it possible for me to enjoy the sweet companionship of a great many persons from whom I should be entirely cut off if I could not talk."

She didn't talk about deaf people. She didn't talk about how at home she felt at Perkins, where most of her peers could speak with their fingers. That's not what anyone there wanted to know about. No one wanted to know about the distance between her and Mabel. It seemed Helen connected with everyone in the Bell family but Mabel. But this was no one's concern. Already, she was being shaped; the priorities of whom it was important to communicate with had been set; her story was being molded to meet hearing expectations.

It wasn't simple, Helen admitted, to learn to speak. And she knew there was work still to do on her voice. But none of that mattered much to Helen. "Remember, no effort that we make to attain something beautiful is ever lost. Sometime, somewhere, somehow we shall find that which we seek. We shall speak, yes, and sing too, as God intended we should speak and sing."

There was a reception afterward, from four to six thirty, attracting more than six hundred people, and though she didn't like crowds, Helen stayed. To listen to the people around her, she placed her hand on the speaker's face—one finger on the nose, one on the upper lip, and one on the chin—to lip-read what they said.

Already, she had become a representative of oralism, and in only a few months she would enter the Cambridge School for Young Ladies, to prepare for her entrance exams for Radcliffe College. She would pass those exams, enter Radcliffe in 1900. In March 1903, she would publish *The Story of My Life,* which she dedicated: "To Alexander Graham Bell[,] who has taught the deaf to speak and enabled the listening ear to hear speech from the Atlantic to the Rockies."

She graduated cum laude from Radcliffe in 1904, becoming the first DeafBlind person to hold a bachelor's degree. By then she was already showing the world that the deaf were not stupid, uneducated, immoral, or soulless. She was inarguably changing the landscape of disability and deafness, but already this was a double-edged sword: her highly curated story was becoming the model of success that all future deaf, blind, and DeafBlind people would be expected to replicate.

As Alec became ever more isolated from the community, and Helen Keller rose as a star, Edward Allen Fay began to publish his genealogical findings, culled from Alec's data and additional survey responses from the deaf community. Chapter by chapter, his book *Marriages of the Deaf in America* appeared in the pages of the *American Annals of the Deaf.*

In the end, Alec had funded a study that found most of his assertions about marriage wanting. To his credit, Alec didn't back away from the information; in 1898, the Volta Bureau underwrote the publication of the study in book form, a genealogical tome.

Deaf-deaf marriages did have more of a likelihood of producing deaf children than hearing-hearing marriages—but deaf people who married hearing people with hereditary deafness in their families were more likely than any group to have deaf children. This confirmed Alec's supposition that the greatest predictor of deaf birth was that of hereditary deafness in families but posed a significant challenge to his solution of deaf people marrying hearing people.

As for Alec's belief that oralism would increase the likelihood that the deaf would marry the hearing, this assertion was also wanting. While Fay's study proved that oral educations correlated, to some extent, with decreased deaf intermarriage, it also suggested that the correlation was small—78 percent of orally educated compared to 86 percent of signing graduates married deaf people. It meant that deaf people who were educated orally still overwhelmingly chose to marry other deaf people.

This finding opened up a new question: Why would deaf people who were educated to enter into the hearing world still choose, so faithfully, to marry other deaf people? The assumption that oralism hinged on was that the hearing world was preferable. But if that were true, why would orally educated people *choose* the deaf community? After all of his data analysis, Fay added one unexpected and unplanned chapter to his book. He titled it, simply, "Happiness."

It's not as though Alec had never considered the question of happiness—he had. When he was asked if, between deaf-deaf marriages or hearing-deaf marriages, one type was happier than the other, he responded that the consensus among principals seemed to be that the deaf-deaf marriages were happier, but he continued that "I know of no data myself from which we can form conclusions." This was the extent of his engagement with that question.

Fay didn't intend to pursue this question, either. It didn't seem measurable—not until he began to notice that the survey returns on deaf marriages often included notes as to divorces or separations. It wasn't a perfect measurement, he knew, but he nonetheless began to tally them up, categorizing them by deaf-deaf marriages, deaf-hearing marriages, and a third category for those who didn't report the hearing status of their spouse. He found that deaf-deaf marriages ended in divorce or separation 2.6 percent of the time; hearing-deaf marriages had more than twice that likelihood, 6.5 percent. According to this data, deaf-deaf marriages appeared more stable, more lasting, and the suggestion was that they were fundamentally happier.

Fay was careful to point out that there were happy marriages of each sort, that this wasn't a sweeping judgment—he himself was a hearing man married to a deaf woman, after all. But he did think these failed marriages were worth considering, as were by extension questions of community and happiness.

Alec had looked to education to determine why the deaf married each other—both whether children attended residential or day schools, and whether their schools were oral or manual. But Fay asks, "Are these influences named the sole, or even the principal, cause of preference of the deaf for marriage with one another?" Since the differences between marriage rates weren't major, he argued that there was something else to look to: "The profounder cause [of deaf people marrying each other] is the deep feeling of fellowship, affinity, kinship, sympathy, which has its roots in the similarity of condition of all the deaf."

It would be another eighty years before the term *deaf culture* would begin to come into widespread use, but its roots were there all along. Deaf marriage was a natural outcome of a people drawn together, and no interventions could stop it: "When opportunity occurs the strong attraction of mutual sympathy draws the deaf together; community of feeling breaks down the barriers that parents and teachers have taken so much pain to erect, sympathy grows into love, and love results in marriage." There was value, Fay suggested, in the community found between the deaf, and within the world of deafness. Marriage was an inevitable outcome, and there was no good reason to prevent it or discourage it. The community gathered around this insistence, pushing Alec further to the fringes. At the 1893 World's Congress of the Deaf, Alec was introduced to a deaf audience as "our friend the enemy."

Alec never saw the deaf community as a valuable one. He believed, against all evidence, that social cures for deafness had to be found by integrating the deaf with the hearing; he believed that deafness was inherently lonely. But it was Alec to whom loneliness clung, even if his greatest effort was to lift it.

Chapter 19

> Jules Verne's books Father always delighted in . . . how often he would speculate on just what a scientific man cast ashore on an uninhabited island, with only a pen knife and his watch, could do to keep alive. With [our houseboat] the Mabel of Beinn Bhreagh on its lonely bit of beach with the private road to it, ending some little distance away, he felt at last that he could find out.
>
> —Daisy Bell

By 1901, the Bells spent most of their time in Nova Scotia. It was there, on a piazza overlooking the Bras d'Or, that Helen and Alec nestled their hands into each other's, communicating through the manual alphabet, discussing fate.

"The more I look at the world," he said, "the more it puzzles me. We are forever moving towards the unexpected." Looking out over the hills of Beinn Bhreagh, he reflected on how much of his reputation had been about the telephone. "And all the time," he said to her, "you know my chief interest is in the education of the deaf."

In the five years since the Flint conference, students taught through the oralist method had leaped from 26 percent to 40 percent. Oralism was on the rise, and Alec wasn't gaining any esteem in the community. How far off course his ship had steered, for want only of a small recalibration early on, for want only of some small flexibility in the belief of his own rightness, some allowance for the input of deaf people. He couldn't see it now. He could only see the vast sea, how still it seemed from where they sat, how lost and indistinguishable each small wave.

Alec transitioned from fate to love. Helen was in her early twenties by then, and Alec wanted to know what she thought of love, as he believed it would soon enough find her.

There was a time when the word *love* was an utter mystery to Helen, when she was just a girl and Anne told Helen that she loved her.

"What is love?" asked Helen.

Anne pointed to Helen's heart. "It is here," she said, and Helen recognized the beating of her heart for the first time.

But she didn't understand. She smelled the violets in Anne's hands and asked, "Is love the sweetness of flowers?"

"No," said Anne.

And so Helen felt the warmth of the sun beating down on them. "Is this not love?" she asked. It was beautiful and warm.

But Anne shook her head. "No," she said.

Days later, when the warm sunlight broke through a cloudy day, Helen asked again, "Is this not love?"

Anne took her hand to explain that warmth wasn't the same as love; love was something more like clouds—something you couldn't touch but which could send down rain, bringing life to flowers.

"You cannot touch love, either," Anne explained, "but you feel the sweetness that it pours through everything. Without love you would not be happy or want to play."

Helen felt the meaning of Anne's words like the sun through the clouds—sudden, pervasive—felt that "there were invisible lines stretched between my spirit and the spirits of others."

Now, on Alec's piazza, she considered it again. "I do think of love," she said, "but it is like a beautiful flower which I may not touch."

They sat in silence for a moment, until his fingers reached again for her palm, "like a tender breath." He told her that she shouldn't think that just because she was DeafBlind that she shouldn't find love. "Heredity is not involved in your case, as it is in so many others."

But Helen insisted that she had love enough: her mother, her teacher. Besides, she said, "I can't imagine a man wanting to marry me. I should think it would seem like marrying a statue."

Perhaps Alec's greatest sin wasn't the structural changes he made to deaf education, nor his opposition to deaf intermarriage. It was something deeper and more pervasive: the way the structures he supported encouraged the deaf to internalize their oppression. Even his most prized pupil had, on some level, succumbed to it. *It would seem like marrying a statue.*

Or maybe Helen was resisting something different, a sense among so many oralists that a deaf person's empowerment had to be found in their communion with the hearing. Maybe she was resisting the idea that a DeafBlind person alone was inherently lonely.

Alec did worry about Helen's loneliness. If Anne were to marry, he said, "There may be a barren stretch in your life when you will be very lonely. . . . If a good man should desire to make you his wife, don't let anyone persuade you to forego that happiness." But when Helen did find love, she wasn't given that opportunity. Though she was in her midthirties, her parents uprooted her when they found out, moving her from Washington, DC, back to Alabama and eventually wielding a shotgun to drive her fiancé, Peter Fagan, away from the house and from Helen.

All her life, the people around Helen sought to keep her, and her legacy, under strict control. It was the only way they thought she could continue to serve in the role of "inspiration." Part of this meant keeping her away from anything that would be perceived by the wide world as political or sexual, and part meant keeping her away from other people who were DeafBlind.

But connections to other deaf, blind, and DeafBlind people remained important anchors in Keller's life, and she kept up a rich correspondence. When Helen was in her twenties, Madame Bertha Galeron, a French

DeafBlind poet, wrote to her to learn about how she read lips. Helen explained the nuances to her, where she placed her fingers to feel the vibrations of sound, how her friends spoke slowly to her so that she could understand, how when her hands got cold the vibrations were harder to feel and she would need to warm them in order to listen again.

Their friendship soon moved on to greater things. They shared the minutiae of their lives, their losses and longings, and their love of writing. Helen wrote to her about *The Story of My Life*, and encouraged her to write her own life story. Galeron sent Helen poems she'd written, which Helen loved. Eventually, Helen translated a few to be published in English. When Galeron wrote to Helen to describe her own history of deafness, Helen "read and re-read [the letter] with love and tears."

"I find myself living in your struggles, sorrows and the passionate joy of hearing again with your hand[s] as if they were my own," she wrote in reply.

Her relationship with hearing and sighted people—especially scientists—was more complicated. They so often sought only to control her and, in line with Mabel's concern that Alec looked upon deaf children merely as cases, Helen was often treated not so much as a human as a specimen.

"I wonder," she wrote, "if any other individual has been so minutely investigated as I have been by physicians, psychologists, physiologists, and neurologists. . . . To scientists I am something to be examined like an aërolite or a sunspot or an atom! . . .

"My scientific tormentors bring all kinds of instruments with long Greek names and strange shapes and appalling ingenuity. Like diabolical genii they check off one's faults and little idiosyncrasies, and record them . . . With mechanical precision they pinch, prick, squeeze, press, sting, and buzz. One counts your breath, another counts your pulse, another tries if you are hot or cold, if you blush, if you know when to cry and laugh, and how fear and anger taste, and how it feels to swing round and round like a large wooden top, and if it is pleasant being an electric battery, and shooting out sparks of lightning—for fun. Resignedly you permit them to bind your wrists with rubber cuffs which they inflate, asking, 'Is it tight or loose?' 'Oh, no,' you answer, 'it doesn't hurt, my arm is quite paralyzed.' . . .

"The tests continue hour after hour, and always a sense of the untrustworthiness of your sensations is borne in upon you. There is a monotonous murmur as the results are read that keep you informed how short you are falling of what was expected of you."

Everywhere, science was taking a sharp turn away from humanity. In 1906, Alec became a member of the American Breeders' Association's Committee on Eugenics, which would become the leading organization in the development of eugenic thought in America. His influence and power in this movement—as in anything he put his name, influence, and money behind—were considerable. By then, he and Mabel owned enough property in Beinn Bhreagh that their land had twelve miles' worth of roads and running expenses of around ten thousand dollars a year. They split their time between Washington, Baddeck, and traveling the world. Alec was known around the globe. By 1881, between telephone stocks, bank and railroad stocks, and US bonds, the Bells became millionaires. From that point on, their money came easily and they spent it freely.

In 1907, George Veditz, president of the National Association of the Deaf, said that the eugenics committee was drafting a legislative bill to restrict marriage among criminals, disabled people, children, "people of plainly incompatible dispositions," the morally defective, consumptives, and the deaf.

"It is evident," said Veditz, "that the one person upon whom we must cast the odium of having hailed the deaf into this category is Dr. Bell, whom his wealth has rendered the most powerful, and his hobby-riding propensity the most subtle, because he comes in the guise of a friend, and, therefore, the most to be feared, enemy of the American deaf, past or present."

By then there was an official committee of the NAD specifically tasked with talking with Alec about the work of this eugenics committee, but when approached regarding the bill, Alec claimed he knew nothing about it. And he wasn't entirely comfortable with his association with the American Breeders' Association, either. He wrote to the association's secretary that he wanted to distance himself from the eugenics efforts: "The Deaf of the country are inclined to hold me responsible for any action of any Committee on Eugenics which touches their interest . . . I am particularly anxious now that my name should appear only in a subordinate position."

Meanwhile, marriage restriction laws were sweeping the nation. In the late-nineteenth and early-twentieth centuries, thirty states passed marriage restrictions for the "unfit." Alec, however, always insisted that legislation be resisted, writing in 1908: "The moment we propose to interfere with the liberty of marriage, we tread upon dangerous ground. . . .

"Among the inalienable rights recognized by the Declaration of Independence are 'life, liberty, and the pursuit of happiness.' The community has no right to interfere with the liberty of the individual and his pursuit of happiness in marriage unless the interests of the community are demonstrably endangered."

He didn't back down on the idea of marriage self-regulation, but he did not believe that the birth of deaf children posed a certain risk of "endangering" the community. It didn't matter. The committee he served on continued to push for legislation—and they went further, too, advocating for sterilization.

In 1905, sterilization legislation inclusive of the deaf passed through Pennsylvania's legislature but was vetoed by the governor. Two years later, in 1907, the first sterilization law passed in Indiana, though it was not inclusive of the deaf. In the next five years, seven more states followed suit, none of them inclusive of the deaf.

By 1909, Charles Davenport, a biologist most famous for his prominent role in the eugenics movement, asked Alec to chair the deafness subcommittee of the eugenics committee. Davenport promised Alec would be able to name the other members. Alec accepted but emphasized that he would not be promoting marriage restriction. For other members, he turned immediately to Edward Allen Fay, explaining that he had accepted chairmanship of the committee "in order to be sure that no recommendation should be made regarding interference with the marriages of the deaf," and begged Fay to join him on the committee: "I feel your attitude . . . is the same as mine. You would not care to serve on the Committee and yet you fear that, without your presence, some action might be taken in reference to marriages of the deaf that would be extremely objectionable."

When Fay turned him down, Alec offered to resign as chair and recommend Fay to the position. When Fay said deaf people should be appointed to the committee, Alec agreed and took Fay's suggestions of two deaf men, James L. Smith and Olof Hanson. Alec again asked that Fay be on the committee and offered to resign completely if that was what Fay wanted. But, he wrote, "I think that the interests of the Deaf demand that one or both of us should serve." There's no record of Fay's response, and Alec continued to chair the committee.

In 1912, ignoring Alec's recommendations, the Committee on Eugenics published a report that included the deaf on a list of groups who should "be eliminated from the human stock," and drafted language for future sterilization laws.

Not long after, in 1915, Helen Keller joined the eugenics movement, writing an open letter to the *New Republic* advocating for a jury of doctors who could make choices about whether babies born with severe disabilities should be permitted to live. Guided in these ideas by Bell, it wasn't until much later in life that Robert Smithdas, a friend and the first DeafBlind person to earn a graduate degree, would explain to Helen the larger complexities of eugenics and his resistance to Alec's ideas.

"He lied," said Helen. "Bell lied."

By 1916, Alec resigned from his work in conjunction with the American Breeders' Association, but efforts to sterilize those determined "unfit" continued. Within ten years, the total number of states to pass sterilization laws hit twenty-five, and the question landed in the US Supreme Court in 1927.

In the landmark case *Buck v. Bell*, John Hendren Bell, superintendent of the Virginia State Colony for Epileptics and Feebleminded, argued for the right to sterilize a woman in his care, eighteen-year-old Carrie Buck, whom he claimed was feeble-minded and promiscuous. The court found in his favor, upholding the state's right to sterilize developmentally disabled people in institutions. This paved the way for even more sterilizations, often including those of the deaf. It's estimated that seventy thousand citizens were forcibly sterilized.

Soon after *Buck v. Bell*, Nazi Germany modeled their early sterilization laws after the United States' legislation. In 1933, Hitler's cabinet passed the Law for the Prevention of Offspring with Hereditary Diseases, which permitted sterilization of disabled people; in 1935, they passed the "Law for the Protection of Hereditary Health of the German Nation," requiring that anyone wishing to marry secure a certificate that stated they were free of hereditary disease; ultimately, teachers at deaf schools complied in delivering deaf children to those who would sterilize them. An estimated 40 percent of the total German deaf population was sterilized during the Nazi era. As time went on, disabled citizens in institutions, including the deaf, were permitted to starve to death, and under the 1939 program "Operation T4," they were systematically murdered.

Alec may have left eugenics behind before much of this came about, but his early involvement remains. It also remains true that his work to promote oralism at all costs laid the groundwork for these eugenic ideas to flourish with the deaf in mind. The values of oralism—that the greatest empowerment for those who are different is found in assimilating to the language and behaviors of the majority—nurtured the seed of eugenic thought. Believing that difference was weakness and that empowerment lay in hiding it away, Alec became convinced that eliminating it entirely was a blessing for all involved.

Alec tapped into subconscious fears of difference that ran rampant in the hearing world, and he harnessed that fear. First had been the fear of freakery, a blurred line between human and animal, a confluence of cultures. Later this morphed into a fear of genetic inheritance, fear of a weakening of the human race. His extreme efforts in support of oralism set up the battles over marriage legislation and sterilization, and his *Memoir upon the Formation of a Deaf Variety of the Human Race* assured that the deaf were central in eugenics discussions from the start.

But more than fear, he tapped into hope—unreasonable, painful, harmful hope. He made promises he could not keep, but in the end, this fact didn't hurt him as much as it hurt the community he purported to serve. Alec promised equality through normality. He promised that the deaf could be taught to speak, and that once they accomplished this feat, the world would be theirs. He held up Mabel as the ultimate example of the success of the method. When it became clear that the method was more difficult than it seemed, he tapped into the genius of Helen Keller. If she could do it, he argued, anyone could.

And it worked. Oralism took hold. Before the Clarke School opened in 1867, all deaf schools in America used ASL. By 1918, approximately 80 percent of deaf students were educated orally. In 1911, Nebraska mandated the oral method in deaf schools; in 1925, Virginia followed suit; in 1929, New Jersey joined them. But legislation almost didn't matter. Approximately 95 percent of deaf children are born to hearing parents, and public opinion in the hearing world had been shaped by Alec. The hearing world wanted speech.

Deaf people knew this. They could see how much their language and culture were in danger. In 1913, George Veditz composed a signed essay, "The Preservation of the Sign Language," funded by the National Association of the Deaf and one of the oldest recordings of the language. It is an astonishing project. In response to the crisis of their endangered language, deaf people had raised $5,000 to create a series of videos—no small task in 1913—to make a record of the language as it existed and to help to preserve it for future generations. Veditz's video was the first of this effort.

Veditz's body begins in a posture of respect. As he talks of the roots of the language, his signs are formal, his movements and face carefully controlled. He bows his head with respect for l'Épée, whom he acknowledges as the father of sign language, and bows his head, too, when speaking of his deaf colleagues. It is the onlooker that he is gesturing toward, across time and space, an invisible audience. Likely we are deaf; at the very least, we sign. His head bows. He does us honor.

He signs with pride for the American deaf, for the fact that in their schools, unlike those in France or Germany, sign language is still allowed. But when he pivots his attention to oralism, his signs slow, the enthusiasm drains from his face and body. He is serious.

He knew that oralism was a major threat—everyone in the community knew—and he called it out directly. "*False prophets* are now appearing, announcing to the public that our American means of teaching the deaf are all wrong." He pauses after this word, for just a moment.

"These men are trying to remove signs from the schoolroom, from the church—from the earth. Our sign language is deteriorating. Old-time masters of sign . . . are rapidly disappearing. In past years, we loved these men, the masters of sign. When they signed to us, we could understand them." This, he signs with tenderness, comfort. When he talks about his community and its roots, the ease is apparent on his body. And when he talks about oralists, his tone shifts. His body is more squared off. His face is, as it has been so far, very controlled, almost guarded.

"As long as we have deaf people on earth, we will have signs," says Veditz. It is only here that you see a smile appear on his shadowy face, the corners of his lips begin to rise almost despite themselves. He continues, "It is my hope that we will all love and guard our beautiful sign language as the *noblest gift* God has given to deaf people."

Alec's most infamous words in the deaf community are these: "We should try ourselves to forget that they are deaf. We should try to teach them to forget that they are deaf." It was what he imagined was possible for all the deaf, that they could slough off that identity and enter invisibly into the hearing world. He believed it was possible, he said, because he saw how Mabel had done it, and later, because Helen Keller had done it.

But Helen Keller didn't forget that she was deaf; instead, she was emboldened by the uniqueness of her perspective. She became one of the most radical thinkers of her time, and this, too, was encouraged by Alec. After she published her memoir, Alec encouraged her to write not only about her own personal sphere but also about the world. She went on to become an outspoken socialist, activist, and feminist; a woman who encouraged men to get into the kitchen in order to learn how economics really worked; she advocated for birth control and an end to poverty; she helped found the American Foundation for the Blind and the American Civil Liberties Union.

And even Mabel, the poster child of a good oralist education, ultimately didn't succeed in the task of forgetting her deafness. When she and Alec were at the end of their lives, in 1921, she finally reckoned candidly with her own deafness. In a letter to her son-in-law, she wrote, "There is one thing I have always put aside because its acceptance involved the acceptance of things which have been my life-long desire to forget or at least ignore—the fact that I am not quite as other people." It was this difference that she had, for years, sought to hide: "I have never been proud of the fact that although totally without hearing I have been able to mix with normal people. Instead, I have striven in every possible way to have the fact forgotten and so to appear so completely normal that I would pass as one."

Even here, Mabel didn't use the word *deaf*; as always, she danced around it as a descriptor of herself. It echoed that legislative hearing from her childhood. Asked if she was deaf, she fumbled. She didn't know what the word meant. She looked up, to the eyes of Miss True, who nodded the answer.

Deaf. She still struggled, on some fundamental level, to identify this way.

She may have been the poster child of oralism, but she was also the poster child of one of its most insidious harms: deep internalized prejudice.

Appearing hearing, appearing "normal," was a guiding tenet of her life; her confession was what she had given up in her attempt to achieve it. Or rather, whom she avoided, a chasm she constructed between herself and other deaf people. "I have helped other things and other people—Montessori, Arts and Crafts, war suffering, anything and everything but the deaf. My deaf cousins were never invited to a dinner party. To say a child was deaf was enough to make me refuse to take any public notice of it." It had gone on for years, for her whole life. "When I was young and struggling for a foothold in society of my natural equals, I could not be nice to other deaf people. It was a case of self-preservation."

As an adult, she made sure her daughters were kept away from deaf people, afraid that they might fall in love and want to marry a deaf person. Her cousin Lena, who was also deaf, liked to go shopping with Mabel, but Mabel never understood this. She felt that alone she could escape notice as a deaf woman, but if she was alongside another deaf person there would be no avoiding the stares of hearing people. "To me it is perfectly awful when a shopgirl turns in perplexity and a little shrinking—oh I have seen it so many times."

The worst offenders, in her mind, were teachers of the deaf. "I was always on the lookout for a little difference in their manner of addressing me, which would reveal the fact that I was a 'case' in their eyes." Yet this was what she'd seen her husband do countless times. He long ago had refused to act any longer as Mabel's teacher, but this dynamic—deaf student–hearing teacher—was the very foundation of their relationship.

"Of all people I hated most was a teacher of the deaf," she wrote. She didn't write about Alec in the letter, not directly.

This letter was a confession, yes, but it was also a pivot. Now she was thinking not of herself, her marriage, her place in society, her daughters or their marriages or their place. Now she was thinking of death, of legacy. "I have now awakened to a realization that I have not done my duty by the memory of the wonderful father and mother who not only did much for their child but so unselfishly labored for the benefit of other deaf children."

She had been thinking of this since Gardiner's death, in 1898. It was then that she went through her father's papers. She learned about her own childhood through her father's eyes and the larger movement her deafness had put into play, the movement to which her husband had dedicated his life. She finally began to reverse her position on his work. "Alec Dear," she wrote, "I never was so in love with the work as now."

When Gertrude died, in 1909, she bequeathed funds to the Clarke Institution for a memorial hall for her husband, to include a chapel. Mabel contributed additional funds to make sure the chapel had the finest finishes—oak paneling, wide windows, and no stained glass, which would hide the view of the valley and the river, the mountains. She added a tablet to the chapel in memory of her mother. The hall was dedicated in 1913.

Now, in the early 1920s, she was thinking again of the Clarke Institution, which she begged her family to cherish "as one of our most precious heirlooms." She wanted her children to give of themselves to the cause of the deaf—specifically, to oralism. She wanted them to give their time and money, and to remember.

Mabel had shown Alec what was possible; Mabel fulfilled his prophecy. But at the end of her life, it was not denial of her deafness that she sustained, but a quiet reckoning. She wrote to her daughter Elsie, "having taught you all my life to forget that I was deaf, I now want you to remember it."

As Mabel aged, she developed cataracts; she lost her ability to read lips clearly. She finally learned the manual alphabet, and she developed the habit of draping her hand over Alec's as he spelled out his thoughts. Though they weren't using sign language in its full linguistic complexity, they were borrowing a feature from a language Alec had worked his whole life to eradicate. This was how they now communicated.

By that point in their lives, Alec finally let go of his work with the deaf. For the last twenty years of his life, he immersed himself in building kites, airplanes, submarines. He still worked on eugenics, publishing occasional articles, but for the most part his life became a quieter one as he grew old.

He had wanted to save the deaf from the struggles they confronted in the hearing world. And yet after devoting a lifetime to liberating deaf people, Alexander Graham Bell went down in deaf history as the culture's great enemy. Within the culture, he is sometimes referred to as the father of audism, of discrimination against the deaf. He wanted to cure the deaf by destroying their language and community. It has never stopped being an unforgivable offense.

Alec never did give up smoking, keeping two pipes around at all times—one to smoke and one to cool. Daisy described him as "a creature of the water and the woods and the night." Toward the end of his life, he could sometimes be seen, after dark, floating in the Bras d'Or, by the little orange light of a cigar.

He was diabetic but never could give up Smithfield hams or apple pie. He gained weight and lost it, and gained it again. He lost strength and gained it, and lost it again. He didn't like doctors, nor following orders.

In his last weeks, Alec was still himself. Maybe more naps, maybe more breakfasts in bed. The only real character difference Mabel noticed was that he was more silent than usual. She would catch him just sitting in his chair, looking into the fire, not talking, not even smoking. Just sitting, deeply still.

And then, in late July 1922, he became weak. The situation quickly declined, but no one panicked. He had been sick before, and Mabel in particular didn't want it to appear that they were worried. It seemed to Mabel that if they didn't call a specialist, if they didn't start making final arrangements, if they behaved as if everything was fine, then her husband would have no reason to think he was dying. And if he had no reason to think otherwise, then he would simply live.

July 22 was the last day he went downstairs, but then his strength left him. He went back up with help, struggling and breathing heavily. He didn't sleep well that night, and though there were several other days that he dressed, he never again went down those steps.

The next day he was up a few times, to his sofa and to his big chair. Mabel thought he might eat, but he didn't. Still, the doctor thought he would pull through. His pulse was weak but steady.

By July 30, he couldn't rise from the sofa. He had no hunger, but at Mabel's pleading, he tried to eat. Daisy and her husband arrived that morning. Alec woke to see them but quickly faded back to sleep. When he woke again, he looked to his wife. "Do you want me to dress?" he asked. "Do you want me to go downstairs?" She didn't. He went back to sleep.

Mabel and Alec slept on the upstairs porch most nights, and his last was no different. Early in the morning of August 1, Mabel noticed he was breathing quickly. She called for the doctor at five thirty in the morning and went back to bed. Alec was sleeping, but his breath was still rapid. He woke for breakfast and ate in bed, propped up by the people around him. He went back to sleep.

When he woke again, he asked for his watch to be fastened to the curtain where he could see it.

The several times Mabel went to check on him, she found him lying there, curtain pushed aside. He didn't say anything or seem to want anything, and so she left.

Around six in the evening, he sat up in bed to give his final dictation to his secretary. When he was finished, everyone left the room but Mabel.

"We have had a happy life together," he said to her. "We have nothing to complain of." In so many ways it was true. His final spoken words to her were "I love you."

As the sun went down, the family gathered in the room again. A lamp was lit, and Alec's son-in-law held a flashlight to his lips so that if Alec spoke Mabel could see his words. But Alec wasn't conscious anymore. Around ten, Mabel lay beside him. He didn't respond, and his breath was so fast Mabel couldn't bear it. She left to lie down in the study.

Around one in the morning Mabel returned. Alec's pulse had faded. He was unconscious; his skin was drawn in loose curtains around his features; his eyes, closed.

"Alec," she said, and he woke, lifted his head, turned to his side and looked at her.

She took his hand. "Don't leave me," she said.

He formed a sign with his fingers, his pointer and middle fingers tapping against his thumb like a pincher, and into her soft palm he said, "No." He wouldn't.

Even then, for a moment, Mabel believed him. She thought he would still pull through, that it was only a matter of when that reservoir of strength within him would fill again, would inflate him with life. She left her hand on top of his. Soon his pulse was too weak to feel at all, but still his fingers made that small tapping movement, that single word—*no, no, no*—into Mabel's waiting palm. Then, in the wake of a last raspy breath, they stilled.

Afterword

In the casket, my grandmother both looked like herself and did not. There was color caked onto her face, finished with powder. She was still and smelled of makeup. She was empty inside.

I was nineteen, but I felt like a child, small and scared. I stood on the other side of the room from her. I had seen death before but this was different. Not like she had been released but like she had been torn asunder. Even though I knew she probably would have died no matter what, I couldn't help but wonder. She was not really *seen* by those doctors. She was not heard. She was not recognized.

At first it was just family in that beige-and-maroon room, standing in small clusters, tense and terse. We didn't really know what to say. What was there to say?

But the whole feeling of the room shifted when the deaf community arrived. They entered with sudden streaks of sunlight, surges of cool autumn air, and the smell of old-fashioned perfumes. They were already engrossed in conversation with one another, their faces trained on one another's movements, colored by the somberness of the situation, and their hands and arms traced out information in the air. They helped one another's walkers through the doors, the crooks of arms linked with the DeafBlind.

We, the family, organized ourselves into a receiving line, and they began to make their way to my grandmother, shaking our hands or hugging us, each in turn. But Charlie, her best friend, walked alone in his pressed suit and fine fedora. He was late-deafened, with only minimal signing abilities, and his face wore the scars from the car accident that deafened him, one eye partially sealed shut. He hugged us each in turn, and he held on tighter than anyone else.

We had known Charlie for years by then. He joined our family for holidays, gifting us beautiful honey hams and organic toothpaste. When we visited my grandmother, he was always there, moving around her apartment as easily as he moved around his own.

In his own apartment, neighboring my grandmother's, sunlight filtered through his stained glass creations, throwing blocks of colorful light everywhere. It was as joyful as his face when he was trying to make her laugh. And after my grandfather died, it was he, perhaps alone in the world, who could do it. He balanced out my grandmother's sternness and inspired smiles of genuine warmth on her face. And she inspired something different in him, too. His signing wasn't very good, and this made him uneasy around some of the fluent signers in the building. But my grandmother was patient with him, teaching him what she could and writing notes back and forth. Sometimes it was less about linguistic communication; he would sit at her kitchen table and talk and talk, and she would make toast and pretend to listen, though she rarely looked at his lips. When he was around she was more grounded, steadier, and there was a release in his body, too, when he was near her. But now, kneeling before her corpse, he looked drained of something vital. I wondered if he'd ever be the same without her.

Her friends and neighbors kept filing before us. Some I recognized from encounters in the elevator, the hallway. Others I'd never seen before. My ASL abilities were minimal, so they each signed the same basic sentences to me.

"I'm so sorry," they'd say, rubbing their fists on their chests, their eyebrows drawn up in quiet concern.

"She was a wonderful woman." Eyes wide and encouraging.

It was hard not to smile a little—most people found my grandmother impossibly difficult, but no matter. We were here to honor her.

"Wonderful woman."

She was.

After Alec's death, plenty of students flourished under oralist methods, but more suffered. Teachers, en masse, could never approximate the ease and talent that Alec had with his students, and so they faced significant challenges in teaching deaf children spoken English. Students floundered. And as they did, crucial windows of language development closed. Early brain development was hindered. Beyond that, these educations were meager of content until children could grasp speech and lip-reading, a task that took up a huge amount of time. Many students struggled in such fundamental ways that they were labeled "oral failures" and graduated "out the back door." Forbidden from signing and inevitably alienated from English, they graduated with no language at all. My grandfather was among them.

My grandfather was born into an all-hearing family. None of them signed. When he was young he went to the Beverly School for the Deaf, which was, like virtually all deaf schools in the 1940s, oral. He arrived at school with no language. He had some gestures he could use to communicate to his family, and a more universal set of gestures to communicate with the world, but they had no grammar. He may have had words, but no ability to build complex sentences, no way to put together a world of ideas.

They tried for years to teach him to speak and lip-read, but with no language to build on, he couldn't access the substance of his education. He was the student Alec never saw, the one oral teachers never wrote to him about. My grandfather lived the situation Albert Ballin warned of: "Oralists in their efforts to suppress the use of signs practically bind the arms of the child, thereby gagging it, so it may not express itself naturally. . . . The attempts to suppress it hinder graceful, upright growth and development, and are worse than no schooling at all." My grandfather didn't begin to learn ASL until his twenties, when he began frequenting deaf clubs.

There, other deaf people began to teach him to sign, and he was able to learn it. It is rare that a person can learn language so late in life, but he managed it. He would still never grasp English, though, not beyond the very basics, and so he never got a promotion at work, he was never taken seriously in the hearing world, and he rarely got credit for his achievements.

My mother remembers the day she realized her father was functionally illiterate. She had just left for college, and she received a card from her parents. On the left-hand side of the card, her mother wrote to her, filling it with sentence after sentence. The right side, she left for her husband, my grandfather. The handwriting was shaky as a child's. He wrote only, *I miss you*. And signed his name, *Harry*.

When I was young, I didn't know any of this. My grandfather was simply my best friend. He would lift me all the way to his shoulders, where I could touch the dusty ceiling and riffle though the baskets on top of the refrigerator, full of pill bottles and jar lids and elastic bands. We would walk around their three-room home like this, him stopping whenever my small hands reached out to run along the rim of a picture frame, press the cold windowpane glass, rearrange the magnets on the refrigerator, one with the ASL sign for *I Love You*, one with two sketched mice, above them the words *Love is You n' Me*.

In their basement was his woodshop, and it was there that he built me a stool three feet high. He stained it a dark brown and gave it to me so that I could be as tall as my grandparents were, could reach the things that they could reach.

I wanted it even taller; I wanted to touch the ceiling, even without his strong, strong arms. "Of course," he said. But then he became sick. He had to slow down his work in the shop. Soon, by the time I was about seven, he couldn't carry me anymore. The last thing he made me was a wooden ball, which I could roll on the floor to him. From his semipermanent place on the couch, he could roll it back.

I never knew my grandfather was illiterate, not during his life. He had all the language I ever needed to communicate with him; he had ASL. But that doesn't mean that the language deprivation he experienced was without impact.

Early language development lays the pathways in the brain that determine not only the ability to gain fluency in other languages later in life but also the abilities of executive function, like cognitive flexibility, problem-solving, and reasoning, as well as social skills like empathy and responsibility.

Wyatte Hall, a deaf population health researcher and a research assistant professor at the University of Rochester medical center, studies language deprivation in the deaf community. He explains that without a language foundation, any person loses a huge amount of their ability to function in this world: "Language is so foundational to who we are as human beings, how we think about the world, how we receive information, how we express ourselves, how— I mean, you can't do any of those things. You're basically cut off from society, in many ways, if you don't have a language foundation."

This happens to varying degrees. A language-deprived person might struggle in things like interpreting social relationships, developing beyond an elementary stage of their understanding of math, and/or establishing basic literacy. Some struggle so completely with language that they can't form complex sentences or anything beyond the most basic interpretations of the world. They do not have an understanding that objects have names or relationships to each other. They are deprived not simply of language but of their ability to understand themselves and the world around them.

While some escape this extreme of a fate, and may be able to pass in public as relatively unscathed, the deficiencies caused by language deprivation still often cause problems in such areas as developing secure interpersonal relationships or finding employment beyond physical labor.

Beyond this, high-stress and high-consequence situations such as encounters with law enforcement, judicial processes, or health-care institutions are often made much worse by lowered abilities for critical problem-solving, social intelligence, and complex language access and expression. Severe instances of over-institutionalization are often seen as a result.

In health-care settings, deaf individuals are made to struggle due to lifelong inaccessibility to information, resulting in high levels of health illiteracy, medication mistakes based on a lack of clear communication, and general neglect. In one example, a study in a psychiatric inpatient unit serving deaf patients found that while 75 percent of deaf inpatients were found to be language dysfluent, only 28 percent had a psychotic disorder (as compared to 88.9 percent of hearing inpatients). Behaviors associated with language dysfluency, such as disorganized patterns of expression, were often misinterpreted as severe psychological symptoms, likely resulting in widespread over-institutionalization.

These barriers extend to prisons, too. Often compounded by a lack of quality interpretation, the problems of language deprivation are seen at every stage of the judicial process. One study of deaf inmates in Texas found that 76 percent were prelingually deafened, implying an increased likelihood of language deprivation, and most were functionally illiterate. It also found that a deaf person was more likely to be submissive to the coercive nature of the police interrogation process, unlikely to be knowledgeable about their right to remain silent or their right to an attorney, and likely to sign paperwork they did not understand, like waivers of rights against a warrantless search or Miranda waivers. While incarcerated deaf people are not officially tracked, HEARD (Helping Educate to Advance the Rights of the Deaf) estimates tens of thousands of incarcerated deaf people.

Language deprivation fundamentally damages individuals within the community and—especially in a communally oriented culture such as deaf culture—damages the community as a whole. At the same time, in a particularly cruel twist, the consequences of language deprivation are still wielded as a weapon to argue against sign language, which is still the most effective tool to combat language deprivation.

In fact, many of the studies that claim deaf children have more behavioral problems, worse education outcomes, and overall more delays than hearing children are measuring the effects not of hearing loss but of language deprivation. A 2017 study that compared hearing children to deaf children with deaf parents (a group of deaf children who had unfettered access to language their whole life), found there were little to no differences in executive function between the two groups. Numerous other studies conducted over the past three decades have shown that neural development in native signers and native speakers is comparable—what actually matters is not what language a child is provided access to, but the fact of that child not facing delays in total language access.

In the hearing world, problems faced by deaf people are often blamed on hearing loss, and so the hearing world seeks solutions to the ear. But it's becoming more and more clear that these problems are more centered on language access than on hearing. The deaf community has been saying this for over a century, but only recently have they gained the access necessary to produce the scientific research to elucidate the problem for a hearing audience. While the deaf community is confronted with the effects of language deprivation nearly daily, the hearing are widely unaware that widespread language deprivation is possible. It's exceedingly rare in the hearing world, and typically requires extraordinary circumstances of abuse and neglect. In his groundbreaking talk on language deprivation, clinical psychiatrist Sanjay Gulati estimates that language deprivation only occurs in about two hearing children a century.

But in the deaf world, the evidence of it is everywhere. It began when the deaf were offered no educations at all, and continued when oralism pushed out sign language–based education. It is still seen today. Language deprivation is inescapable within the community, invisible outside of it.

In 1907, when George Veditz said that Alec was "the most to be feared enemy of the American deaf, past and present," even he hadn't predicted the way "future" would be added to that statement. Bell's name and legacy are still being utilized today to suppress ASL.

At the Clarke School, after my grandmother's death, I peered into classrooms where children now had cochlear implants and teachers covered their mouths as they spoke, preventing the children from relying on lip-reading to understand. This was modern-day oralism, cochlear-implant education, which in most cases continues to suppress ASL. Many in the hearing world view implants as a blanket "good," allowing the deaf to access more of the sounds of the world around them; however, many in the deaf world still balk at the ASL-forbidding educations that are promoted alongside the technology.

Though implants have long been touted as a miracle cure for deafness, the truth is that implants do not restore hearing as it is understood by hearing people. The sounds received through an implant still take years of education and training to interpret because the brain needs to essentially create new pathways between auditory input and meaning, and success rates for this rebuilding vary widely. Like oralism, success relies on these years of work, and it doesn't work for everyone. The work, for children, is often accomplished through "Listening and Spoken Language" (LSL) educations.

Like oralism, these educations have Bell's fingerprints all over them. In the years after Bell's death, the American Association to Promote the Teaching of Speech to the Deaf morphed into the Alexander Graham Bell Association for the Deaf. As such, it was an early and influential supporter of implants and LSL educations. LSL is oralism rebranded for implants—like oralism, LSL educations forbid the use of sign language and focus on skills for listening and speaking. This is despite the fact that implanted children who learn sign language alongside spoken language show better results in language acquisition and development.

Oralism and LSL form one unbroken line, and in ASL, Bell-the-man and Bell-the-organization share a name, *AG Bell*, representing the whole of the set of ideas supported by his legacy. Many cite this set of ideas—oralism and LSL educations—as the sustaining force behind language deprivation. They are the forces that systematically withhold unfettered language access from deaf children.

If Bell was hated by the signing deaf community before, his organization's active role in promoting implants and LSL educations ripped open all of oralism's old wounds. As research into language deprivation is proliferating, it's become more and more clear—scientifically clear—that deaf children who have free language access their whole lives perform at similar levels of language development as hearing children. The deaf children who struggle are those who cannot easily access language, and this includes children with cochlear implants. Cochlear implants can be a helpful tool for deaf people, but they are not a cure. The most consistent liberator of the deaf is not cochlear implants, but total language access. Signing is language deprivation's preventative medicine. And yet, less than 8 percent of deaf children grow up with regular sign language access.

I began this book trying to understand the roots of the oppressive forces that had dictated so many parts of my grandparents' lives, and that search took me in more directions than I could have imagined. I used to believe Bell was simply a villain—his very name inspired nothing but rage within me. And so most surprising to me, as I learned more about him, were the ways that Bell revolutionized many things in the deaf world for the better. He helped fund the educations of countless deaf children, helped to open deaf schools across the nation, in many ways helped to advocate for the idea that the deaf were the same as anyone else.

But oralism was painful enough and powerful enough to negate any of those accomplishments. Oralism hobbled the early intellectual development of generations of deaf children and led to massive under-education and alienation of deaf children from deaf communities. Bell's greatest sin wasn't trying something new. His mistake was turning his back on what the deaf community had to say, and then on the evidence that pushed against him, even when it showed up in his own school.

In a 2017 paper about deaf epistemology and Albert Ballin's anti-oralist *The Deaf Mute Howls,* Octavian E. Robinson and Jonathan Henner explain that even today, "the privileging of scientific knowledge produced by non-deaf people, coupled with emphasis on sound-based languages . . . and ableism, allows myths and ignorance to persist despite two centuries of publications by deaf people advocating sign language pedagogical methods for deaf children." Deaf knowledge, shaped by centuries of experience, has always held strong to sign language; yet studies on hearing and development habitually neglect to draw upon this knowledge.

There is a whole array of elements that can confound the results of studies on hearing, many of which are common knowledge in the deaf world. Deaf exclusion means that these same errors are made again and again, creating a flawed body of data that repeatedly skews results toward hearing values. Historically, these errors have included failing to take into account the way bilingualism affects early language development (making monolingual spoken-language speakers and bilingual users of a spoken and a signed language incompatible for assessment); the differences between measuring children who learn sign language later in childhood and those who have had language access their whole childhood; and the value judgments inherent in weighing spoken language quality over deeper demonstrations of language expression and comprehension.

At the same time, signing has been accepted by the hearing community as good for hearing babies, and "baby signs" are frequently taught to hearing infants. No one worries that this will interfere with these children's ability to learn spoken language. Yet the deaf community is still embroiled in a battle to overturn oralism's insistence that sign language is bad for the language development of deaf babies. As the deaf community has been pointing out for decades, there's a devastating irony here.

The hearing world has appropriated the deaf community's greatest asset, while continuing to deny it to the deaf themselves.

Even in Bell's time, the deaf knew the importance of sign language for deaf education, but Bell didn't value that knowledge. He was a scientist at heart and thrilled at a new hypothesis, at the execution of a great experiment. That experiment failed, but he still couldn't believe he was wrong. First Bell turned his back on deaf cultural knowledge, then he turned his back on his own evidence. In doing this, he turned his back on generations of deaf children.

The deaf community was right when they said that there were many deaf children for whom oralism simply didn't work, and oralism never found a solution for these children. Instead of graduating prepared to enter and be embraced by the hearing world, many of them were prepared for no world at all. My grandfather was the hidden risk of oralism: the student who was promised everything and given nothing.

The sign for *low* is a simple one: the hand shape for the letter *L*, slowly sinking from the center of the signing space to its bottom. There are a million ways to use this sign; I've seen it in all manner of conversation. But this sign has a weight for me—it is the sign my great-aunt Rita used, over and over again, to describe her position in relation to deaf people who could speak. *Low*. They were higher, on a different level. In the signing space, they towered above her. She? Low. She had had her power beaten out of her, every time a teacher slapped her signing hands, every time they hid her signing parents behind closed doors, every time they punished her for not being able to hear. She was a product of the Clarke School, which she attended in the 1930s and '40s, about seventy years after the school's founding was inspired by Mabel.

Even in her eighties, she had not recovered from oralist ideas. In the years after my grandmother's death, when I set out to understand oralism, I began to interview Rita. By then, my ASL was very rough, and so my mother interpreted these interviews. Every time we sat down to talk about her experiences, my aunt seemed bewildered. "Why do you want to write about this?" she'd want to know. "Why would anybody read it? Why would anyone care?"

But the stories themselves seemed to strengthen her. By the interview's end, when I'd ask if there was anything she'd like to add, she always began, "I don't know if it's important, but . . . ," and she would tell more.

During these interviews, my mother would always want to share the story of the dictionary. It was 1965 when it happened. My mother was in college, and some people at Gallaudet University had written a dictionary of ASL. It was a wild and mystical thing, widely scoffed at by the deaf community. A dictionary carried weight, carried implications that had been educated out of so many deaf people, and so some were slow and cautious in believing in something so beautiful: a dictionary meant ASL was a language. If ASL was a language, it couldn't be minimized anymore.

My mother, a college student at the time, was elated. She was given the dictionary, a tan-colored hardback, its title pressed into its cover and inked in royal blue. It was a first edition that could only be purchased at Gallaudet University's bookstore, and she brought it home.

"It's a language," she said, "a real language."

Her uncle Frank rolled his eyes at his naive niece. "No, it's not," he said. "It's just gestures."

At the time of the interviews, in the early 2000s, this was something we could all laugh at. Of course ASL was a language; by then it had passed into fact. But even so, Rita's face would shadow over when she saw the story. She could laugh now, she knew she had to laugh now, but it was as though she still couldn't believe it. After all these years, it still seemed suspect that her language was real, that it had power, and maybe it was hard for her to believe that she was powerful, too.

But in her community, her power was undeniable. For her and her husband's fiftieth wedding anniversary, nearly a hundred people filled a local restaurant, the Porthole, to celebrate, most of them deaf. They laughed and told stories, and gave toasts that my mother interpreted into English for those us who knew only that language. By then, that included me; ASL had slipped and slipped from me for years.

Friends remembered when Rita and Don met, when Rita would lie to her mother to sneak out to see him. Rita's eyes were glittery listening to these stories; glittery when she looked at her husband, with his walrus mustache and apple cheeks; glittery with the deaf applause that erupted, fingers twinkling in the air. She and Don moved through the crowd easily, signing with their friends, all of them using what Rita had once claimed to be "just home signs," the dignity of the language having been beaten out of her. Everyone seemed to know these "home signs," though. It was language, nothing less.

The ASL dictionary was part of a new turn in deaf history. In the 1960s the invention of the TTY, a device that uses phone lines to send text, finally gave the deaf access to the telephone. In the 1970s, ASL began to come back into schoolrooms under the "total communication" method, which advocated for both speech and ASL to be used in deaf education. It was problematic in execution but still an improvement over pure oralism. By the end of the 1980s, the Deaf Power movement was in full swing: Marlee Matlin won an Oscar for her role in *Children of a Lesser God*; Linda Bove was exposing a generation of children to ASL on *Sesame Street*, and more and more culturally positive books and videos were published.

The great apex of the movement, the Deaf President Now protest, took place in 1988, when the trustees of Gallaudet University appointed yet another hearing person to the role of president of the university. Instead of making an official announcement to the people who had gathered on campus on the day of the board's decision, the community found out by the university's PR staff handing out flyers. They were incensed.

Students spilled out of the campus and onto Florida Avenue. There, Gary Olsen, president of the National Association of the Deaf, looked out over the crowd—students crestfallen, furious, shocked—and began to sign: "We're standing here in the middle of the street, in the middle of the cold, and we don't know what to do. The Board is over there at the Mayflower Hotel drinking fancy wine, eating delicious food, and laughing at us." And so they marched to the hotel.

There, the chair of the board, Jane Spilman, begged them to give the new president a chance. Instead, the students went back to campus and began to plan. Led by students Bridgetta Bourne, Jerry Covell, Greg Hlibok and Tim Rarus, students, staff, and faculty at Gallaudet staged an eight-day protest, garnering national media attention and forcefully rejecting over a century of hearing paternalism that had ruled the school. Because of the protests, the trustees withdrew their choice and appointed, instead, I. King Jordan, a deaf man. He became the first deaf president of the university.

Oliver Sacks, who was present at the protests, wrote, "Will deaf people at Gallaudet, and the deaf community at large, indeed find the opportunities they seek? Will we, the hearing, allow them these opportunities? Allow them to be themselves, a unique culture in our midst, yet admit them as co-equals, to every sphere of activity?" These were, and remain, genuine and pressing questions.

In person, Bell was intensely humble—rarely did anyone ever report feeling stupid around Bell, even though he was surely one of the great minds of his time. In person, he listened and he conversed and he signed. It was never, in practice, his goal to make others feel small or worthless, but because he divorced his ideas from people—and spent more and more time focused on the ideas—that's exactly what he ended up doing. After the failure of his school, his work with the deaf never had anything more to do with the deaf themselves. As he veered into eugenics, he was almost entirely cut off. His work was done in dusty archives, alone, alone, alone, alone.

A century later, my grandmother lay on a hospital bed, sapped of any power she had wrested out of her life. My grandmother was vastly different from her sister, but their personalities were, in part, shaped by the same oppression. There was something beneath her fierceness, a feeling like she wasn't sure she believed it, as though it were an act.

In her life, I pretended I didn't see it. It was the least I could do. I believed in her fierceness, after all. I did not believe in the imagined worthlessness that it may have been an attempt to cover. I still believe these things, but in her death, I began to see their interplay. As the fierceness betrayed her, that other feeling—the one her sister revealed so easily—began to shine through. Lying for three days alone in the hospital, struggling for language, battling just to be recognized . . . it was as though it were all a confirmation of the thing she struggled so hard to deny: that she did not have worth, that she was low, an *L* sinking slowly, a frown. Like it was only a matter of height or hierarchy, a simple fact.

In a way, I fell victim to this thinking, too. I knew it was wrong, but in the year after my grandmother's death, I still kept asking myself, over and over again, *What if oralism had worked?* I saw my grandmother weak, and I wanted her strong; I saw her desperate, and I wanted her soothed.

Maybe this was the greatest risk of my having understood Bell flatly as a villain—doing so made it that much more difficult to admit the villain that I carried inside of myself.

Every savior narrative begins with compassionate human instinct: you see what you perceive to be suffering and you seek to alleviate it. According to Darwin, it is the noblest part of our nature, and yet it can lead to the darkest places, the greatest violence and violation. Saving can become more important than recognizing; fixing more important than loving. This was Bell's greatest mistake. He got so caught up in the feeling of sympathy that he never veered into empathy; his imagination didn't stretch enough to understand, or even ask, what the deaf themselves wanted; and in the face of rejection and failure, he pushed harder, dug down deeper, argued ever more from a single, fixed point. He aspired to something so big that it would be perceived as a miracle—but his focus on the miracle came to subsume the work of observation, attention, empathy. He let the saving get the better of him.

When I was a child I understood that, as a carpenter, my grandfather fixed things. I set aside broken things to bring to him, and he fit them back together again, made them whole. I piled broken things at my grandfather's feet.

After his death, I still brought him my broken things—broken dreams, broken promises, things I broke and things that broke within me. In the dark I always asked him, *What would you do?*

As a child I knew the answers plainly: he would be strong and steady; he would be good to anyone; he would always make time to play; he would never yell.

As I grew up, the stories became more complicated: he would swing his hammer all day long, even as he knew that his supervisors would never promote him to foreman; he would jump off a lobster boat into the stormy ocean to save the life of a young man; he would go fishing in the foggy dawn, get stoned and dance on the beach at night; he would find himself some peace.

My grandfather had all the answers I ever needed, but for this question—*How did Bell go so wrong?*—my grandfather *was* the answer. Bell turned his back on my grandfather; he let him grow up without language. There were ways my grandfather could never be in the world, not because of his deafness, but because he was language deprived.

Even though my grandmother didn't suffer the same fate as my grandfather, she was also the answer to my question about Bell. She grew up in the deaf world, in a deaf family, and so she always had access to language. When she arrived at school, she endured a different kind of trauma: she was punished for using the only language she knew. Every day, oralists taught her that her way of being was shameful.

In Bell's time, the oralists already knew these outcomes. They were observed by the deaf community, whose members did not fail to share their opinions and experiences. But Bell didn't take the time or the effort or the humility to be quiet and listen. He used his power to do the thing he prized most: to speak.

My grandparents deserved none of what came to them from Bell. My grandfather deserved an education; he deserved language. And my grandmother deserved to follow her dreams, go to college, die with dignity. But the damage of oralism, of Bell's legacy, ensured they had none of this.

My grandparents spent their whole lives trying to be recognized in this hearing world, but their greatest liberation was the deaf world, where they never had to engage in this struggle.

I loved their world. Our minds and bodies worked differently because of it—ASL molded our minds to process and express information spatially, visually, and temporally—and it carried with it the sweet safety of my grandparents' love. But it's also true that I was a guest in that world, and there is a part of me now that can fall into the trap of peeking into this world as a hearing person and seeing only the beauty. This may seem like a positive thing—the deaf world, from the outside, can appear fascinating, beautiful, magical—but this way of thinking is dangerous, too. Focusing on the beauty can distract from the fact that this world is necessary. If we see it only as beautiful, then in arguments about money, innovation, and power, it is understood as superfluous. Deafness is still not seen as a culture, and so efforts toward its erasure, and the erasure of its language and people, are seen as benevolence instead of forms of systematic ethnocide. Though access to this community should be assured to deaf people as a basic human right, the truth is that the community still has to spend seemingly limitless time, money, and energy fighting for its own sustained existence.

Bell was able to get away with what he did because he was protected by a power structure that did not value deaf experience or deaf voices. Bell thought he was bringing the deaf and the hearing closer together, but in refusing to allow for deaf agency, he drove us further apart. His ideas continue to drive us further apart.

When my grandmother died, we were not close. In my teenage years, I had grown distant from her and her culture. By the time she was in the hospital, I'd mostly forgotten her language. It was within this distance that Bell's ideas began to slip into my thinking. I wanted her close, and I wanted her powerful, but some part of me wanted all this effort to come from her. The power conferred to me by the systems we moved in was palpable, and it drove a wedge between us. My thoughts during her death—my wondering if her deafness was a problem—were a betrayal.

I wish I could have been with her in death the way we'd all been during my grandfather's last years. When my grandfather was still strong, he lifted me up; he showed me all the things he could see from where he stood. *Love is You n' Me.* And then he built me a stool, so I could stand tall without him. Soon he was too sick, too weak to lift me anymore, and he lay on the couch while I sat on the floor beside him. But we were not *low*. We were simply together, and stronger for it.

My grandmother sat beside me in the recliner, her arms and fingers chattering with my grandfather and me. If someone was looking away, her hand would pat the air beside us, in the periphery of our vision, and we would turn to listen. My grandmother told us what she could see out the window: the birds in the feeder, swaying the branches of the sapling it dangled from. The movement was something she could capture perfectly in ASL—the pace of wingbeat, the tottering, a branch flexing, a playful chickadee dangling upside-down for a moment before righting itself. Meanwhile, my grandfather and I took turns coloring that little wooden ball he'd made me in his workshop, in the most beautiful crayon colors we could find.

As the sun went down, my grandmother looked down at us with tenderness, the light catching her fingers as she signed. She was so hard in life, and this tenderness was reserved only for the sacred spaces in which she could move freely. Here, it spilled out of her: her eyes soft, her face receiving. I sat on the floor, and my grandfather lay on that couch, and we rolled that ball back and forth to each other, in the pulsing light of my grandmother's words, until the day he died.

Acknowledgments

First, thank you to my agent, Farley Chase. I am so impossibly grateful for you. From the beginning I could see how hard you were going to work for this book, but I didn't see how compassionate you would be, how whenever you caught the edge of worry on my voice, you would tell me about a particular scene in the book that you loved. You always managed to assure me that things were going to be okay, and then worked to make that true. Knowing you were there, steadily supporting this effort, was constant relief.

The first editor of this book, Jonathan Cox, your suggestions always felt like they were coming from someplace inside of my own vision for the book that I could not access myself. You were patient and insightful always. You helped me move every abstract thought closer to its human core and were as exacting with sentences as you were with ideas. I am so lucky and grateful to have had your eyes and mind at work on this book.

Megan Hogan and Priscilla Painton inherited this book and helped it cross the finish line. Megan, thank you for your meticulous notes, the care you gave every sentence, for asking so many tiny questions, and later for answering so many tiny questions. Priscilla, thank you for your expertise and clear vision for how to bring this book out into the world. I am so deeply grateful to you both for the enthusiasm you brought to this project. Thank you to Jonathan Karp, for believing in this book early on. Thank you also to Elise Ringo and Cat Boyd for your thoughts and work and sensitivity in bringing this book to its audiences. To the assistants, copy editors, production managers, layout and cover designers, and everyone else who worked quietly behind the scenes at Simon & Schuster, thank you.

At Scribe UK, thank you to Philip Gwyn Jones for early interest in this book and Molly Slight for your care as we brought it out into the world, and the whole rest of the team at Scribe.

Several institutions offered critical support to this effort. Thank you to the Edward Albee Foundation and the Massachusetts Historical Society, two early supporters of this project, who gave me encouragement when I sorely needed it, and who helped me begin to see myself as both an artist and a historian. Thank you to the Blue Mountain Center, sanctuary masquerading as residency. I owe BMC for so many friendships, so much revelation, so much recalibration. Thank you to all my fellow residents over the years and to the people who make that place run. Thank you especially to Ben Strader, Nica Horvitz, and Harriet Barlow, for creating a space of so much tranquility and community.

The Kluge Center at the Library of Congress was my headquarters for nearly a year, and that fellowship was one of the most staggering, stunning experiences of my life. The resources, time, and community they provided allowed this book to become what I always wanted it to be. Thank you to John Haskell, who was at the helm. Thanks to Staci Jones, Mike Stratmoen, and Dan Turello. Hugs to Travis Hensley and Emily Coccia for all the gossipy lunches when I definitely needed some giggles. Thank you to all my fellow fellows, especially those who were there with me for the long haul: Sarah Bell, Kristen Shedd, Davide Ceriani, Luke Harlow, Christy Chapin, Andrew McGee, and Ivan Sigal. And thank you to Catherine Able-Thomas, who played quiet defense when I was in the depths of this manuscript, making sure no one pressured me to emerge from my little cave.

A special thanks to the librarians in the Manuscripts Division of the Library of Congress, all of whom helped day in and day out to preserve and provide the resources that made this book possible: Frederick Augustyn,

Loretta Deaver, Joseph Jackson, Patrick Kerwin, Edith Sandler, and Lewis Wyman. Extra thanks to Lara Szypszak, who scanned materials for me after Covid-19 closed library to the public, and to Bruce Kirby, who was my pal long before I had any fancy fellowship to legitimize me, when I was just a weirdo who showed up on a yearly pilgrimage with a bleary look and a busted camera.

Thank you also to those in the Prints and Photographs Division, especially Ryan Brubacher, who helped me find photos during the Covid-19 closure.

At the Smithsonian Museum of American History: thank you to Katherine Ott, for putting together a fine archive of disability ephemera and for putting me in contact with other historians at the museum. Science historians Hal Wallace and Barney Finn patiently took me through the collections of early telegraphy and telephony, explaining the ins and outs of the way these machines functioned, as well as the larger cultural context in which they were invented. Thank you especially to Barney, for patient readings of the invention sections of the book, and for helping me better understand and represent the mechanics of the telephone.

The very early days of this book were made possible by the collections at Historic Northampton, the Nielson Library at Smith College, and the archives at the Clarke School for the Deaf. Later research was conducted on-site at Harvard University's Houghton Library and Schlesinger Library, and the Gallaudet University Archives. Thank you to all of the librarians and archivists who supported the work that I was able to conduct there. Thank you to the collectors and enablers of digital collections as well, which made so much of this research possible: the Boston Public Library, the American Foundation of the Blind, and the Perkins School for the Blind. The Disability History Museum had a tremendous online collection long before it was common, and for that I am very grateful.

Many thinkers, writers, and scholars helped to support this book, but some were more hands-on than others. Thank you to: Wyatte Hall, who was the first person to really crystallize the idea of language deprivation syndrome for me, and years later, for reviewing portions of the book that pertain to language deprivation and its effects. Camille Bromley offered edits on a story I wrote, many years ago, that caused me to rethink and recalibrate my approach to this book as well—it was critical and I am beyond grateful. Kenny DeHaan watched "The Preservation of the Sign Language" with me, and filled in the nuances of Veditz's language that I couldn't catch myself; helped troubleshoot and brainstorm a variety of problems, and generally offered reassurance and encouragement. Luke Harlow talked with me about this era of history and added greater context to it, and offered encouragement and pointers for navigating academic-leaning systems as an academic-system-adverse person. Chris Jon Heuer provided friendship and patience as I relearned ASL, and always has keen insight into the collisions between deaf and hearing cultures. Richard McGann shared memories of Robert Smithdas. Ernie Freeberg talked with me about writing history and technology in ways that remain narratively engaging and looked over and responded to sections of this book. Christine Sun Kim granted us the use of her art on the cover of the book—I am a huge fan, and it is such an honor. John Lee Clarke was an incredibly patient and enthusiastic teacher, and helped to inform the way I wrote about Laura Bridgman and Helen Keller. Brian Greenwald offered support, manuscript feedback, research pointers, and his own research, which I relied on heavily and which has shaped my thinking in more ways than I can count. And he is maybe the one person on earth with whom I can gossip about the tiniest minutiae of Bell's life.

This project accompanied me through two schools: Hampshire College and the University of Pittsburgh. Thank you to the teachers who fundamentally expanded my sense of what I could do with this work: Michael Lesy, Ernie Alleva, Joel Lovell, and Jeanne Marie Laskas. And to the teachers who taught me the nitty-gritty details of how to research, edit, and write every day: Peter Trachtenberg, Jean Carr, Ellie Siegel, Lynne Hanley, Susan Tracy, Will Ryan, and Annie Boutelle. Before any of that, Tom Gotsill and Anna Gado set the groundwork and believed in me.

Nearly two decades ago, my first foray into writing this book was interviewing several graduates of the Clarke School. They sat down with me, sometimes many times over, to share their experiences of oralism. Thank you to George Balsley, Rodney Kunath, Martha Morrell, Malcolm Nash, Roger Retting, and Gini Spaulding. These conversations fundamentally altered the way I thought of this history, rendering it much more complex than I had previously understood. This book would not have existed without their perspectives complicating my preexisting beliefs.

A few people stand out as absolutely critical to the book: Melina Ragazas, who took an early draft of the full manuscript and helped shape it from an amorphous blob into something resembling a narrative, thus saving me from total humiliation before my editor. I cannot even begin to express my gratitude.

Jeff Mansfield, one of the smartest people I know, played so many parts in the making of this book: Fellow fellow at the LOC, a historian of deaf culture, commiserator over the writing and research processes, and most importantly, a friend. I can't imagine what this book would be without our many conversations while I was frantically getting this down on the page.

Cristina Hartmann, I am incredibly grateful for our conversations in the late stages of editing this book. Your feedback, both on the page and off, was essential. I am so grateful for the way you were able to take problems of content and representation and help to troubleshoot them through the perspective of a writer. Thank you especially for your support and perspective as I worked through the sensitivity read edits. I'm so deeply grateful for your presence in my life as I've navigated it all.

My sensitivity reader, who remains anonymous, you transformed this book in profound ways and I am deeply indebted to your patient, straightforward and thorough reading of this manuscript. It changed the book, and it also changed me.

My community of writers and artists buoys me week after week, year after year. There are too many to name here but some people stand out as being there when I needed it the most. Thank you to Tim Maddocks, Jill and Ad Yeomans, Geeta Kothari, Amy Whipple, Eddie Martinez, Becky Cole, Maggie Jones, and Sejal Shah, whose friendship, advice, and support were critical to this book and my relative sanity while writing and editing it. Thank you, Renee Prymus, Gavin Jenkins, and Marissa Landrigan, for feedback at critical moments along the way. To my intentions group, Tim Maddocks (again), Tyler McAndrew, and Rachel Wilkinson, and the offshoot morning writing meetups, thank you for helping me stay on track (and get out of bed) as the pandemic raged. To the folks who made up the Pittsburgh Writers Collective, I'm so grateful for all the hours I spent working and writing beside you.

To those in my chosen family, whose love and support has meant so much to me, who have guided me and kept me steady, thank you. I could not have written this if not for the escapes I had with you—to the ocean, lakes, forests, mountains, rooftops, porches, and stoops. You each helped hold me together when I felt I could not hold myself together. Maureen Cotton, you have been my close friend longer than anyone, and know me in ways no one else does. Thank you for reflecting that back to me with love and compassion. Amrith Prabhu, so much laughter and so much of everything else. Thank you for always being there and for waiting at the bottom of the stairs. Rachelle Escamilla, for joining me in burn-it-to-the-ground furies, for all the laughter, and so many magnificent meals. Sam Milford, I'm so grateful for the grounding you bring to my life, tempered with such creativity, joy and laughter. Martin Doppelt, who also deserves thanks for translating several letters between Helen Keller and Bertha Galeron, thank you for always listening, and for always being ready with a cat video when I didn't want to talk. Jonathan Smucker, you have taught me so much about what is possible. Thank you for every hike, every canoe ride. Kate Weed, lifestyle coach extraordinaire, thank you for sharing in so much of the early process of recalibrating my life toward writing, and for permitting my use of your internet during all hours as I did. I know it was absurd. Miriam Devlin, so many walks. Thank you for writing such weird, beautiful sentences and for always being there for escapes. Thank you, too, for patiently tutoring me in the science of telegraphy and early telephony. Kristian Kaseman, you saw me through some of my worst, when I could not see through my own anxiety, and you always made me laugh again. Thank you for all the quiet.

I would be such a fundamentally different person if I hadn't had the opportunity to witness my family's deaf culture. Thank you to Rose McCarthy, Harry McCarthy, Don Cyr, Frank Burke, Leo Burke, Elise Burke, and Lena Burke. This book owes a special debt to my great-aunt, Rita Cyr, who told me hard stories raked through with shame and sadness. Thank you, too, to Michael Cyr, for helping to fill out my memories of our deaf family.

To my favorite uncles—C.F. Booth, John Keefe, Michael Booth (in *alphabetical* order)—thank you for supporting the strange life choices I've made, and for all the love and whiskey. My sisters Lena Booth and Mary Kay Zukowski (somehow I was blessed with sisters who don't like to read), thanks for keeping me in check, and for putting up with all the gifted books. Here's another! Thank you to my mother, Barrie Booth, for her fierceness in the face of audism, and for teaching me to question power and fight injustice. And to my father, Tom Booth, for encouraging me from a young age to record my observations, both on tiny cassette tapes and in journals, with curiosity and gentleness.

Notes

MANUSCRIPT AND PHOTOGRAPH SOURCES

The Alexander Graham Bell Family Papers, Manuscript Division, Library of Congress, Washington, DC
Boston Pictorial Archive, Print Collections, Boston Public Library
Archives, Clarke School for the Deaf (now at the University of Massachusetts, Amherst)
Disability History Museum, Buffalo, New York
Archives and Deaf Collections, Gallaudet University, Washington, DC
Grosvenor Collection of Alexander Graham Bell Photographs, Prints and Photographs Division, Library of Congress, Washington, DC
Houghton Library, Harvard University, Cambridge, Massachusetts
Helen Keller Archive, American Foundation for the Blind, Arlington, Virginia
Helen Keller Papers, Schlesinger Library, Radcliffe Institute for Advanced Study, Harvard University, Cambridge, Massachusetts
Massachusetts Historical Society, Boston, Massachusetts
National Museum of American History, Washington, DC
Historic Northampton, Northampton, Massachusetts
Old Boston Photograph Collection, Print Collections, Boston Public Library
Archives, Perkins School for the Blind, Watertown, Massachusetts
T. H. Gallaudet and Edward Miner Gallaudet Papers, Manuscript Division, Library of Congress, Washington, DC

Unless otherwise noted, all diaries, journals, notebooks, and letters are from the Alexander Graham Bell Family Papers, Manuscript Division, Library of Congress.

On Miracles

ix *In 1679*: Paul T. Jaeger, Cynthia Ann Bowman, ***Disability Matters: Legal and Pedagogical Issues of Disability in Education*** (Westport, CT: Bergin & Garvey,

2002), 3; Alexander Graham Bell, "Historical Notes," 34, Alexander Graham Bell Family Papers, Manuscript Division, Library of Congress.

Prologue

10 *But not everyone*: Alexander Graham Bell, "The Utility of Signs in the Instruction of the Deaf," *The Educator* 5 (1898): 38–44.

15 *I wanted to know*: Albert Ballin, *The Deaf Mute Howls* (Washington, DC: Gallaudet University Press, 1998), 40–41.

PART 1

Chapter 1

19 *Mere voice*: Alexander Bell, *The Tongue: A Poem* (London: W. J. Cleaver, 1846), 41.

19 *In 1863*: Alexander Graham Bell, "Making a Talking Machine," 1–7.

19 *His father, Alexander Melville*: Alexander Melville Bell, *Visible Speech: A New Fact Demonstrated* (London: Hamilton, Adams, 1865), 25. This alphabet, in Melville's view, would far surpass other attempts, like Karl Richard Lepsius's "Standard Alphabet," Max Müller's "Physiological Alphabet," even Alexander Ellis's "Universal Writing."

19 *Melville spent much of his time*: Charlotte Gray, *Reluctant Genius: Alexander Graham Bell and the Passion for Invention* (New York: Arcade, 2011), 9.

20 *Recently, Melville's research*: Bell was at times unsure of the timing of this visit, whether or not it happened before he constructed his own machine, though it seems likely it happened before.

21 *Through high school*: Alexander Graham Bell, "Moral Education in Childhood," *Beinn Bhreagh Recorder*, November 22, 1909.

21 *Where Aleck's default*: Robert V. Bruce, *Bell: Alexander Graham Bell and the Conquest of Solitude* (Boston: Little, Brown, 1973), 308.

21 *The Bells had a camera*: Photos found within the Grosvenor Collection of Alexander Graham Bell Photographs, Prints and Photographs Division, Library of Congress.

21 *At the encouragement of his father*: Alexander Graham Bell, "Notes from Early Life," 2–3.

22 *In the attic*: Ibid., 3.

23 *"convert the unlettered . . ."*: Charles David Bell and Alexander Melville Bell,

Visible Speech: The Science of Universal Alphabetics; or Self-Interpreting Physiological Letters, for the Writing of All Languages in One Alphabet (London: William Mullan & Son), ix.

23 *it allowed an*: Alexander Melville Bell, *Visible Speech: A New Fact Demonstrated*, 13.

24 *The mid-nineteenth century*: Jonathan Rée, *I See a Voice: Deafness, Language and the Senses—A Philosophical History* (New York: Metropolitan Books, 1999), 213–14.

24 *"in no higher respect"*: Alexander Melville Bell, "Public Reading: The Causes of Its Defects; and the Certain Means for Their Removal," *The Athenæum Journal of Literature, Science, and the Fine Arts*, no. 1776 (November 9, 1861): 616–17.

24 *Philosopher Johann Gottfried Herder*: Rée, *I See a Voice*, 67.

24 *In ancient Greece*: Rée, *I See a Voice*, 94–95.

25 *He was aware*: Alexander Melville Bell, *Visible Speech: The Science of Universal Alphabetics*, 8.

25 *It was because of*: Bruce, *Bell*, 5.

25 *But the effect*: Thomas L. Hankins and Robert J. Silverman, *Instruments and the Imagination* (Princeton, NJ: Princeton University Press, 1995), 214.

26 *Now she rested*: Bruce, *Bell*, 21–22.

26 *Before Melville had started*: Fred DeLand, "Brief Biographical Notes," 18.

26 *Eliza was also*: Alexander Melville Bell, letter to Elsie and Marian, undated; Eliza Symonds Bell, diary, 1850.

26 *Melville had fallen*: Alexander Melville Bell, letter to Elsie and Marian, undated.

27 *As they fell*: Eliza Symonds Bell, diary, 1850.

27 *For Eliza, sketching*: Eliza Symonds Bell, "On Observations."

27 *She'd lost her hearing*: DeLand, "Brief Biographical Notes," 10.

28 *"played the Scottish melodies"*: Alexander Melville Bell, letter to Elsie and Marian, undated.

30 *"our triumph and happiness"*: Alexander Graham Bell, "Making a Talking Machine," 5.

31 *At first his flights*: Alexander Graham Bell, "Notes from Early Life," 8.

31 *And so he began*: Ibid.

31 *"I spent many"*: Gray, *Reluctant Genius*, 13.

31 *"Dear Guide"*: Alexander Graham Bell, letter to Alexander Melville Bell, March 1, 1864.

32 *In the summer*: Alexander Melville Bell, *Visible Speech: A New Fact Demonstrated*, 25–29.

33 *Aleck looked at*: Ibid.

33 *Even with these*: Ibid; Bruce, *Bell*, 42–43.

33 *To drum up*: Helen Elmira Waite, *Make a Joyful Sound: The Romance of Mabel Hubbard and Alexander Graham Bell* (Philadelphia: Macrae Smith, 1961), 57.

33 *One offered*: Catherine Mackenzie, *Alexander Graham Bell: The Man Who Contracted Space* (Boston, New York: Houghton Mifflin, 1928), 32.

34 *He was still hoping*: The British government officially passed on it on February 12, 1867.

Chapter 2

35 *"I put it to any"*: Gardiner G. Hubbard, *The Story of the Rise of the Oral Method in America, as Told in the Writings of the Late Hon. Gardiner G. Hubbard* (Washington, DC: Press of W. F. Roberts, 1898), 32.

35 *The year before*: There is a lot of disagreement about how old Mabel was when she lost her hearing. The story often went that she was either four or five, but later in life Mabel found documents that cited the date of her illness a year later. Possibly she was billed as younger to make her attainment of speech seem more impressive.

35 *Twenty-one years old*: P. T. Barnum [author unknown but suspected to be Barnum], *Sketch of the Life, Personal Appearance, Character and Manners of Charles S. Stratton, the Man in Miniature, Known as General Tom Thumb, and His Wife, Lavinia Warren Stratton; Including the History of Their Courtship and Marriage, with Some Account of Remarkable Dwarfs, Giants, & Other Human Phenomena, of Ancient and Modern Times, and Songs Given at Their Public Levees,* pamphlet, Disability History Museum.

36 *It must have been*: Ibid.

36 *The fever trembled*: Robert McCurdy, letter to Miss Evelyn McCurdy, *Beinn Bhreagh Recorder* 23, 6. Alexander Graham Bell Family Papers, Manuscript Division, Library of Congress.

36 *Still, Mabel remained alert*: Ibid.

37 *Even before Mabel*: Mabel Bell, memories written in Mabel's handwriting. Alexander Graham Bell Family Papers, Manuscript Division, Library of Congress.

37 *Gertrude could enter*: Robert McCurdy, letter to Miss Evelyn McCurdy, *Beinn Bhreagh Recorder* 23, 6–7. Alexander Graham Bell Family Papers, Manuscript Division, Library of Congress.

37 *After about a week*: Ibid.

37 *Gertrude sent for*: Ibid.

37 *And then the day*: Waite, *Make a Joyful Sound*, 20–27.

38 *While Mabel had been*: Barnum, *Sketch of the Life*.

38 *In the image*: There are several cartes de visite of Lavinia Warren, but I believe this is the only one that fits the date.

38 *"Little lady"*: Waite, *Make a Joyful Sound*, 20.

39 *Aside from this question*: Mary True, "Fragmentary Recollections of Mary Hatch True," *Beinn Bhreagh Recorder* 24, 1920, 90. Alexander Graham Bell Family Papers, Manuscript Division, Library of Congress.

39 *Mabel had deaf cousins*: Though there is mention of Mabel's deaf cousins, there is almost no accessible information on who they were, how they went deaf, how they were educated, or any other details beyond very minimal mention of one cousin named Lena.

39 *Inside its brick walls*: *The Forty-Eighth Annual Report of the Directors of the American Asylum at Hartford, for the Education and Instruction of the Deaf and Dumb* (Hartford, CT: Press of Case, Lockwood, 1864), 7–8; Hubbard, *Oral Method in America*, 21–22.

40 *Turner, however*: Hubbard, *Oral Method in America*, 21–22.

41 *Thomas followed*: Christopher Krentz, ed., *A Mighty Change: An Anthology of Deaf American Writing, 1816–1864* (Washington, DC: Gallaudet University Press, 2000), 6.

41 *"condemned all their* life": Ibid., 9–10.

42 *"I had it is true"*: Ibid., 10.

43 *The son of a*: Bruce, *Bell*, 83.

44 *Sitting in Turner's office*: Hubbard, *Oral Method in America*, 5.

45 *At Gardiner's meeting*: Waite, *Make a Joyful Sound*, 18–19.

45 *Back home*: True, "Fragmentary Recollections of Mary Hatch True," 65.

46 *There was a deaf man*: Horace Mann, "Seventh Annual Report of the Secretary of the Board of Education," *The Common School Journal* 6 (no. 5) (March 1844): 79.

47 *On the south edge*: Waite, *Make a Joyful Sound*, 22.

47 *Howe was a man*: Ibid., 23.

48 *Unlike William Turner*: Waite, *Make a Joyful Sound*, 23.

48 *"He was an Angel"*: Elisabeth Gitter, *The Imprisoned Guest: Samuel Howe and Laura Bridgman, The Original Deaf-Blind Girl* (New York: Farrar, Straus and Giroux, 2001), 51.

49 *Precursor to braille*: First known as Howe Type, later as Boston Line Type.

49 *The word* normal: Lennard J. Davis, *Enforcing Normalcy: Disability, Deafness, and the Body* (London: Verso, 1995), 24.

49 *Laura, this woman*: Dorothy Herrmann, *Helen Keller: A Life* (New York: Alfred A. Knopf, 1998), 20.

50 *Howe explained*: Waite, *Make a Joyful Sound*, 24.
50 *Gertrude described him*: Mabel Hubbard Bell, diary, February 16, 1884.
50 *Like her husband*: Waite, *Make a Joyful Sound*, 23.
51 *Gertrude's was a prayer*: Ibid., 25.
51 *Mabel's lessons began*: Ibid., 27.
51 *If she wanted*: Ibid.
51 *That was the idea*: Ibid.
51 *Gertrude worked*: True, "Fragmentary Recollections of Mary Hatch True," 82, 90.
52 *As the months*: Tony Foster, *The Sound and the Silence: The Private Lives of Mabel and Alexander Graham Bell* (Halifax, Nova Scotia: Nimbus Publishing, 1996), 35.
52 *In March 1864*: Hubbard, *Oral Method in America*, 6.
52 *However, members*: Ibid., 24.
52 *"we are satisfied"*: Hubbard, *Oral Method in America, 25.*
53 *In the fall*: True, "Fragmentary Recollections of Mary Hatch True," 62–63.
53 *Eight years old*: Ibid.
53 *Now Mabel*: Waite, *Make a Joyful Sound*, 31.
53 *Mabel liked*: True, "Fragmentary Recollections of Mary Hatch True," 71.
54 *Miss True was guided*: Mabel Hubbard Bell, memories written on the occasion of Miss True's death.
54 *In Mabel's mouth*: Ibid., 75.
54 *They sat*: Ibid., 82.
54 *Mabel worked*: Ibid., 72.
54 *"the mere getting"*: Mabel Hubbard Bell, memories written on the occasion of Miss True's death.
55 *"their recovery of articulation"*: *Deaf-Mute Education in Massachusetts: Report of the Joint Special Committee of the Legislature of 1867, on the Education of Deaf-Mutes: with An Appendix containing Evidence, Arguments, Letters, Etc., submitted to the Committee* (Boston: Wright & Potter, 1867), 61.
55 *"I think our friends"*: Ibid., 69.
55 *For Howe*: Hubbard, *Oral Method in America*, 32–33.
56 *On January 31*: Waite, *Make a Joyful Sound*, 38; *Deaf-Mute Education in Massachusetts*, 43–45. This scene is based largely on Waite's retelling, which supposes that Gardiner was present at this hearing, which is in conflict with the legislative reports. Much of Waite's book was based on interview material, so it may be that this was a story based in fact that morphed in its retellings over time. I have edited Gardiner out of the scene, based on the information in the reports.

Chapter 3

58 *"We are satisfied"*: Alexander Melville Bell, *Visible Speech: A New Fact Demonstrated*, 23.

58 *His roommate woke*: Bruce, *Bell*, 45.

58 *And when his roommate*: Ibid.

59 *The next night*: Alexander Graham Bell, letter to Alexander Melville Bell, November 24, 1865.

60 *"I can fully sympathize"*: Alexander Melville Bell, letter to Alexander Graham Bell, February 14, 1866.

60 *"I implore you"*: Eliza Symonds Bell, letter to Alexander Graham Bell, May 24, 1866.

60 *"Young birds"*: Eliza Symonds Bell, letter to Alexander Graham Bell, March 17, 1866.

60 *In Elgin*: Mackenzie, *Bell*, 35; Bruce, *Bell*, 47.

61 *"Edward died today"*: Alexander Graham Bell, scribbling diary, 1867.

61 *By the time he*: Eliza Symonds Bell, account books.

61 *Melville gave up*: R. Cassily, "Information on the Bell Family, Requested by Mr. H. G. Long of the Bell Telephone of Canada"; Bruce, *Bell*, 53.

61 *Melville, meanwhile*: Mackenzie, *Bell*, 27.

62 *reported with great enthusiasm*: "Professor Alex. Graham Bell Seeking to Make Animals Talk," *Halifax Herald*, reprinted in *Beinn Bhreagh Recorder*, November 11, 1910, 345. Alexander Graham Bell Family Papers, Manuscript Division, Library of Congress.

62 *Aleck encouraged the dog*: Bruce, *Bell*, 56.

62 *He felt that he*: Alexander Graham Bell, "Reminiscences of the Early Days of Speech Training," *Volta Review* 14, no. 8 (December 1912): 579–81. Reprinted in *Beinn Bhreagh Recorder* XII, December 23, 1912, 170–75. Alexander Graham Bell Family Papers, Manuscript Division, Library of Congress.

62 *On May 21*: Alexander Graham Bell, journal (while teaching at Hull's), 1–2.

63 *He taught the students*: Ibid., 2–3.

63 *Aleck knew nothing*: Alexander Graham Bell, untitled article draft, August 15, 1904.

63 *When the children described*: Alexander Graham Bell, journal (while teaching at Hull's), 5.

63 *Aleck loved this work*: Ibid., 8.

63 *By June 4*: Ibid., 10.

64 *With meaning came*: Ibid.

65 *In 1868*: This year is an educated guess.

65 *For months*: Melly Bell, letter to Alexander Melville Bell, undated.

65 *Melly missed his brother*: Melly Bell, letter to Alexander Graham Bell, April 1, 1870. Emphasis in the original.

66 *He was drawn to*: Melly Bell, letter to Alexander Melville Bell, undated.

66 *As the winter wore*: Ibid.

66 *"The quantity of pure bile"*: Ibid.

66 *"Let me give"*: Melly Bell, letter to Alexander Melville Bell, July 25, 1869 [date penciled on the envelope].

66 *"Another day of fog"*: Bruce, *Bell*, 63.

67 *"Can it be?"*: Melly Bell, letter to Alexander Graham Bell, May 24, 1870.

67 *Two days later*: Alexander Graham Bell, letter to Alexander Melville Bell, May 25, 1870; Alexander Melville Bell, letter to Alexander Graham Bell, May 24, 1870.

67 *"With pillows"*: Alexander Melville Bell, letter to Alexander Graham Bell, May 24, 1870.

67 *"death had no terrors"*: Alexander Melville Bell, letter to Alexander Graham Bell, May 28, 1870.

67 *In his grief*: Bruce, *Bell*, 76.

68 *"Our earthly hopes"*: Alexander Melville Bell, letter to Alexander Graham Bell, May 28, 1870.

68 *Aleck, however*: Cassily, "Information on the Bell Family, Requested by Mr. H. G. Long of the Bell Telephone of Canada," 4.

68 *The doctor gave him*: "Condensed Biographical Notes Concerning Alexander Graham Bell (1847–1922) a Citizen of the United States," 6. This story of Bell being diagnosed with consumption is hard to verify. Bell has told the story, but the timeline is incredibly cramped for such a diagnosis, and there's no reference to it in the letters of the time. That said, there is an urgency about Aleck's health around that time that may suggest such a thing happened.

68 *"We have resolved"*: Alexander Graham Bell, letter to Alexander Melville Bell, May 30, 1870.

68 *But Melville wasn't*: Ibid.

69 *"he has a stock"*: Alexander Graham Bell, letter to Alexander Melville Bell and Eliza Symonds Bell, June 1867. Emphasis in the original.

69 "*I* do *wish you*": Bruce, *Bell*, 66.

69 *"I will treat myself"*: Melly Bell, letter to Alexander Graham Bell, July 2, 1869 [date stamped on the envelope].

70 *They held hands*: Bruce, *Bell*, 67.

70 *"You know that I cannot"*: Alexander Graham Bell, letter to Alexander Melville and Eliza Symonds Bell, June 5 1870.

70 *In Edinburgh, Aleck went to work*: Ibid., June 19, 1870.

70 *"You don't really think"*: Eliza Symonds Bell, letter to Alexander Graham Bell, June 20, 1870.

71 *Aleck sold Melly's piano*: Alexander Graham Bell, "Reminiscences of the Early Days of Speech Training."

Chapter 4

72 *The public looked upon*: "Unpublished Biography of Dexter King," 19. Alexander Graham Bell Family Papers, box 83, Manuscript Division, Library of Congress.

72 *When they arrived*: Alexander Graham Bell, letter to Mabel Hubbard Bell, September 26, 1875; Eliza Symonds Bell, letter to Alexander Graham Bell, April 8, 1875.

72 *He found it*: Alexander Graham Bell, letter to Mabel Hubbard Bell, September 26, 1875.

72 *This Canadian forest*: Letter from Alexander Graham Bell to Mabel Hubbard Bell, September 26, 1875.

74 *This was where King*: "Unpublished Biography of Dexter King." Originally, King had hoped that his school would be a branch of the Clarke School, but found it too difficult.

74 *Clarke was exactly twice*: *Fourth Annual Report of the Clarke Institution for Deaf-Mutes at Northampton, Mass., for the Year Ending February 1, 1871* (Boston: Wright & Potter, 1871), 27. The annual cost for a residential student at Clarke was $350. *The Annual Report of the Directors and Officers of the American Asylum, at Hartford, for the Education and Instruction of the Deaf and Dumb* (Hartford, CT: Wiley, Waterman & Eaton, Steam Book and [indecipherable] Printers, 1871). The annual cost for a residential student at the American Asylum was $175.

75 *"I speak with"*: Ida Montgomery, "The Practical Value of Articulation," *American Annals of the Deaf and Dumb* 15, no. 3 (1870): 131–36.

75 *the amount of money*: *Deaf-Mute Education in Massachusetts*, 60. The cost to educate in 1867 was $200/student, up from $182; only $100 was covered by the state, down from $115 in 1830.

75 *Montgomery wrote*: Montgomery, "Practical Value of Articulation."

75 *To make this vision*: "Unpublished Biography on Dexter King"; Dexter King, letter to Alexander M. Bell, March 7, 1871. Emphasis in the original.

76 *He suggested to King*: "Unpublished Biography on Dexter King."

76 *In addition, King*: Ibid.

76 *Boston smelled of*: Boston Pictorial Archive, Boston Public Library Print Collections; Alexander Graham Bell, letter to Eliza and Alexander Melville Bell, April 16, 1871.

76 *In the next few days*: Alexander Graham Bell, letter to Eliza and Alexander Melville Bell, April 16, 1871. The man who gave him the tour was Thomas Hill, president of Harvard.

77 *He also witnessed*: Alexander Graham Bell, letter to family, April 9, 1871.

77 *The building*: "Unpublished Biography on Dexter King."

78 *Now there were*: Alexander Graham Bell, letter to Eliza and Alexander Melville Bell, April 16, 1871.

78 *Aleck admired*: Ibid.; Alexander Graham Bell, letter to Eliza and Alexander Melville Bell, April 9, 1871.

79 *Later, with*: Alexander Graham Bell, letter to Eliza and Alexander Melville Bell, April 16, 1871.

79 *"I may say"*: Ibid.

79 *"First impressions"*: Alexander Graham Bell, letter to Eliza and Alexander Melville Bell, April 16, 1871.

80 *Though Aleck may have*: Alexander Graham Bell to parents, April 16, 1871. The student's name was Isabella Flagg.

80 *He spent that evening*: Ibid.

81 *He met with*: Ibid.

81 *When it was*: Ibid.

81 *Then, when Aleck*: Ibid.

81 *The audience*: Ibid.

82 *When another man*: Ibid.; Alexander Graham Bell, letter to Alexander Melville and Eliza Symonds Bell, April 16, 1871; "The Final Benefaction of an Eminent Methodist," Alexander Graham Bell Family Papers, box 83, Manuscript Division, Library of Congress.

82 *"The public looked upon"*: "Unpublished Biography on Dexter King."

83 *At the American Asylum*: Bruce, *Bell*, 87.

83 *Aleck thought*: Alexander Graham Bell, "Reminiscences of the Early Days of Speech Training."

83 *Out of politeness*: Ibid.

Chapter 5

84 *"Sometimes on a dark day"*: Mabel Hubbard Bell, notebook number 10, 1922.

85 *She wrote letters*: Mabel Hubbard Bell, letter to Gertrude Hubbard, May 26, 1873.

85 *In Washington, Gardiner*: Bruce, *Bell*, 126; A. Edward Evenson, *The Telephone Patent Conspiracy of 1876: The Elisha Gray–Alexander Bell Controversy and Its Many Players* (Jefferson, NC: McFarland, 2000), 18–25.

86 *He also believed*: Ibid.

86 *For years now*: Ibid.

86 *"How are you getting on"*: Mabel Hubbard Bell, letter to Gardiner Greene Hubbard, February 2, 1874.

87 *"I am almost afraid"*: Lilias M. Toward, *Mabel Bell: Alexander's Silent Partner* (Wreck Cove, Nova Scotia: Breton Books, 1996), 17.

87 *"quack doctor"*: Mabel Hubbard Bell, diary, January 6, 1879.

87 *None of this was*: Ibid.

88 *At the time, Mabel*: Mabel Hubbard Bell, letter to Gertrude Hubbard, undated, c. 1873. Emphasis in the original.

88 *On its face*: Mabel Hubbard Bell, letter to Gertrude Hubbard, undated, c. winter of 1873–1874.

90 *Meanwhile, their students'*: *Sixth Annual Report of the Clarke Institution for Deaf-Mutes at Northampton, Mass., for the Year Ending February 1, 1873* (Boston: Wright & Potter, 1873), 7.

91 *He visited an invention*: The phonautograph he visited was Leon Scott's design, which had since been improved by Charles Morey.

91 *MIT also housed*: The manometric capsule was Rudolph Koenig's design.

91 *From the vowel sound*: Alexander Graham Bell, letter to Alexander Melville Bell, April 1874.

91 *And so Aleck began*: Alexander Graham Bell, letter to Eliza Symonds Bell, April 4, 1874. Emphasis in original.

92 *black walnut bookcases*: True, "Fragmentary Recollections of Mary Hatch True," 62.

92 *"He was so entertaining"*: Mabel Hubbard Bell, letter to Gardiner Greene Hubbard, February 2, 1874.

92 *Mabel read lips*: Alexander Graham Bell, "Reminiscences of the Early Days of Speech Training."

92 *"Since there is"*: Alexander Graham Bell, "Visible Speech" (speech, Boston Society of the Arts, April 1874).

92 *"there will soon be"*: Alexander Graham Bell, "Visible Speech" (speech, Massachusetts Institute of Technology, November 4, 1874).

93 *Mabel looked forward*: Mabel Hubbard Bell, letter to Gardiner Greene Hubbard, December 7, 1874.

93 *The weeks went by*: Mabel Hubbard Bell, letter to Gardiner Greene Hubbard, December 21, 1874.

95 *Into the evening*: Mackenzie, *Bell*, 61; Alexander Graham Bell, *Upon a Method of Teaching Language to a Very Young Congenitally Deaf Child* (Washington, DC: Gibson Brothers, 1883); Waite, *Make a Joyful Sound*, 82.

95 *"When deaf-mutes are"*: Alexander Graham Bell, "Visible Speech" (speech, Boston Society of the Arts, April 1874).

96 *"If we can find"*: Alexander Graham Bell, letter to Eliza and Alexander Melville Bell, May 6, 1874.

96 *As Aleck spent*: Gray, *Reluctant Genius*, 46.

96 *On February 2, 1874*: Mabel Hubbard Bell, letter to Gardiner Greene Hubbard, February 2, 1874.

97 *But she didn't* want: Mabel Hubbard Bell, letter to Gertrude Hubbard, February 3, 1874.

98 *But Aleck said*: Mabel Hubbard Bell, letter to Gardiner Greene Hubbard, February 2, 1874.

98 *The type of weather*: Bruce, *Bell*, 304; Mackenzie, *Bell*, 229; Marian (Daisy) Bell, memories of Alexander Graham Bell, written on the train, July 4, 1926.

98 *Back home*: Mabel Hubbard Bell, letter to Gertrude Hubbard, February 3, 1874; Toward, *Mabel Bell*, 22–23; Bruce, *Bell*, 101; Waite, *Make a Joyful Sound*, 85. Waite's is the only mention of him taking her all the way home, but is also based in part on oral history, so it may include information there is no other record of.

98 *Nearly a year later*: Mabel Hubbard Bell, diary, January 6, 1879.

99 *"How well I remember"*: Gray, *Reluctant Genius*, 80.

99 *Aleck played for*: Mackenzie, *Bell*, 76. This pause isn't consistent across sources, but is present in Mackenzie, which is drawn in part from oral histories not available elsewhere.

100 *"The potentialities of the telegraph"*: Bruce, *Bell*, 127.

100 *But now a gaunt*: Bruce, *Bell*, 125–27; Alexander Graham Bell, letter to Eliza and Alexander Melville Bell, October 23, 1874; *The Bell Telephone: The Deposition of Alexander Graham Bell in the Suit Brought by the United States to Annul the Bell Patents* (Boston: American Bell Telephone Company, 1908), 955, Massachusetts Historical Society; Mackenzie, *Bell*, 76; Gray, *Reluctant Genius*, 78–79.

Chapter 6

102 *"I am like a man"*: Bruce, *Bell*, 149.

102 *Early in the winter*: Thomas Watson, *Exploring Life: The Autobiography of Thomas A. Watson* (New York: D. Appleton, 1926), 55. Watson said this occurred in "early 1974" (*Exploring Life*, 54), but this seems inconsistent with other facts and accounts, which place it in the winter of 1974–1975.

102 *Thomas had no idea*: Ibid.

102 *In Williams's shop*: Thomas Watson, "How Bell Invented the Telephone" (speech, Annual Meeting of the American Institute of Electrical Engineers, New York, May 18, 1915); Watson, *Exploring Life*, 31–35.

103 *"Is red hot glass"*: Watson, *Exploring Life*, 48–49.

103 *His father was*: Ibid., 3.

104 *His plan was*: Watson, *Exploring Life*, 56.

104 *Thomas Sanders was less thrilled*: Alexander Graham Bell, letter to Eliza Symonds and Alexander Melville Bell, October 23, 1874.

105 *In a* Chicago Tribune *article*: "Telegraphing Tunes," *Chicago Daily Tribune*, July 12, 1874, 4.

105 *The news was alarming*: Gardiner Greene Hubbard, letter to Alexander Graham Bell, November 19, 1874.

105 *"He has the advantage"*: Alexander Graham Bell, letter to Eliza Symonds and Alexander Melville Bell, November 23, 1874.

106 *"Work as if"*: Eliza Symonds Bell, letter to Alexander Graham Bell, November 30, 1874.

106 *By January 1875*: Bruce, *Bell*, 135; Watson, *Exploring Life*, 56–57.

106 *At Mrs. Sanders's home*: Ibid., 59.

106 *More than anything*: Ibid.

106 *Thomas, too, lived*: Ibid., 58.

107 *As they walked*: Ibid.

107 It was many and: Charles David Bell and Alexander Melville Bell, *Bell's Standard Elocutionist* (London: William Mullan & Son, 1878), 282. Note: It's not known what poems exactly he recited, but it is likely they would have come from those collected in *Bell's Standard Elocutionist.*

107 *On one walk*: Mackenzie, *Bell*, 81.

107 *Aleck spread*: Ibid.

107 *Thomas listened patiently*: Ibid.

107 *Thomas was more measured*: Ibid.

108 *They strung wires*: Watson, *Exploring Life,* 60–61.

108 *They could successfully*: Evenson, *Telephone Patent Conspiracy*, 52; Watson, "How Bell Invented the Telephone."

108 *"The chief result"*: Watson, *Exploring Life*, 57.

109 *By late January*: Alexander Graham Bell, letter to Gardiner Greene Hubbard, January 26, 1875.

109 *They were, in fact*: Alexander Graham Bell, *The Multiple Telegraph: Invented by A. Graham Bell* (Boston: Franklin Press: Rand, Avery, & Co. 1876), 18. Alexander Graham Bell Family Papers, Manuscript Division, Library of Congress.

109 *Gray was on the forefront*: Bruce, *Bell*, 137–39; Christopher Beauchamp, *Invented by Law: Alexander Graham Bell and the Patent That Changed America* (Cambridge, MA: Harvard University Press, 2015), 39–40.

109 *Two nights before Aleck*: Alexander Graham Bell, letter to Eliza and Alexander Melville Bell, February 13, 1875. Emphasis in original.

110 *While he was in DC*: Ibid.

110 *"My visit to the Smithsonian"*: Ibid.

110 *"The signals were"*: Grovesnor, *Alexander Graham Bell*, 56.

111 *That afternoon*: Alexander Graham Bell, letter to Eliza and Alexander Melville Bell, March 22, 1875.

111 *"You seem to think"*: Alexander Graham Bell, letter to Eliza and Alexander Melville Bell, March 18, 1875.

111 *"The worst of it is"*: Eliza Symonds Bell, letter to Alexander Graham Bell, April 10, 1875.

112 *He came, and he*: Gray, *Reluctant Genius*, 89.

112 *Mabel wore a silk*: Ibid.

112 *Aleck, exhausted*: Bruce, *Bell*, 143.

112 *"Watson, we are on"*: Watson, *Exploring Life*, 61.

113 *He began with*: Alexander Graham Bell, notebook of lessons with Mabel Hubbard.

113 *"If you run your finger"*: Ibid.

114 *It was Aleck's job*: Watson, "How Bell Invented the Telephone," May 18, 1915.

114 *In theory, when*: Watson, *Exploring Life*, 73.

115 *"vivid shop language"*: Watson, "How Bell Invented the Telephone," May 18, 1915.

115 *This was different*: Watson, "How Bell Invented the Telephone."

116 *At first he*: *The Bell Telephone*, 59.

117 *"A meager result"*: Watson, "How Bell Invented the Telephone," 7.

117 *"I am like a"*: Bruce, *Bell*, 149.

117 *And all the while*: Eliza Symonds Bell, letter to Alexander Graham Bell, June 20, 1875.

118 *While Aleck quickly*: Watson, "How Bell Invented the Telephone," 8.
118 *There were problems*: Bruce, *Bell*, 148–49.
118 *Aleck wrote to*: Alexander Graham Bell, letter to Gertrude McCurdy Hubbard, June 24, 1875.
119 *Gertrude was clear*: Alexander Graham Bell, diary, June 25, 1875.
119 *"My Pride told me"*: Alexander Graham Bell, letter to Mabel Hubbard Bell, August 8, 1875. Emphasis in original.
119 *As the commencement speeches*: Alexander Graham Bell, letter to Mabel Hubbard Bell, August 8, 1875.
120 *the way he used words*: Marian (Daisy) Bell, memories of Alexander Graham Bell, June 13, 1932.

Chapter 7

121 *"I feel so misty"*: Mabel Hubbard Bell, letter to Gertrude Hubbard, 1875.
121 *A few weeks later*: Alexander Graham Bell, "The Pre-Commercial Period of the Telephone," *Beinn Bhreagh Recorder* 9, Alexander Graham Bell Family Papers, Manuscript Division, Library of Congress.
121 *The same day*: Bruce, *Bell*, 152.
121 *"She is beautiful"*: Alexander Graham Bell, letter to Eliza and Alexander Melville Bell, June 30, 1875.
121 *Eliza was unenthusiastic*: Eliza Symonds Bell, letter to Alexander Graham Bell, July 4, 1875.
122 *late July*: Late July is an estimate based on the date of Aleck and Gertrude's correspondence and the length of time Aleck was sleepless.
122 *Mabel wrote to her mother*: Mabel Hubbard Bell, letter to Gertrude Hubbard, 1875.
122 *"Help me please"*: Ibid.
122 *By the time*: Bruce, *Bell*, 153.
123 *On Saturday, August*: Alexander Graham Bell, diary, 1875.
123 *The next morning*: Ibid.
123 *Aleck lay down*: Ibid.
123 *Mabel had no hearing*: Mabel Hubbard Bell, diary, January 6, 1879.
123 *He knew, once he*: Alexander Graham Bell, diary, 1875.
124 *Aleck was grateful*: Alexander Graham Bell, letter to Mabel Hubbard Bell, August 8, 1875.
124 *By Monday*: Alexander Graham Bell, diary, 1875.
124 *Aleck had begun*: Ibid.
125 *There, Mary greeted*: Ibid.

125 *He hailed a ride*: Ibid.

125 *And he would have to*: Ibid.

125 *In Salem, Mrs. Sanders*: Alexander Graham Bell, letter to Alexander Melville Bell, August 31, 1875.

125 *Aleck didn't go*: Alexander Graham Bell, diary, 1875.

126 *He slept until*: Ibid.

126 *When he finally did*: Alexander Graham Bell, letter to Eliza Symonds Bell, August 18, 1875.

126 *On Eliza's use*: Ibid.

127 *"It is the duty"*: Alexander Graham Bell, *Marriage: An Address to the Deaf*, 3rd ed. (Washington, DC: Sanders Printing Office, 1898), 4.

127 *"Now that I have done"*: Alexander Graham Bell, letter to Mabel Hubbard Bell, August 10, 1875.

127 *Days later he wrote*: Alexander Graham Bell, letter to Gardiner Greene Hubbard, August 14, 1875.

128 *The magneto could power*: Ibid.

128 *"Thank you very much"*: Mabel Hubbard Bell, letter to Alexander Graham Bell, August 15, 1875.

129 *Now this shifted*: Alexander Graham Bell, letter to Mabel Hubbard Bell, December 12, 1885; Alexander Graham Bell, letter to Gardiner Greene Hubbard, August 14, 1875.

129 *Now Gertrude kept him*: Gertrude Hubbard, letter to Alexander Melville Bell, August 19, 1875.

129 *Mabel came home*: Gertrude Hubbard, letter to Alexander Graham Bell, August 20, 1875.

129 *"She is well but"*: Ibid.

129 *Aleck thought he should*: Alexander Graham Bell, letter to Gertrude Hubbard, August 24, 1875.

130 *Gertrude finally surrendered*: Gertrude Hubbard, letter to Alexander Graham Bell, August 25, 1875.

130 *Mabel met him*: Alexander Graham Bell, diary, August 26, 1875.

130 *"Perhaps it is best"*: Mabel Hubbard Bell, letter to Alexander Graham Bell, August 30, 1875.

131 *He didn't tell them*: Alexander Graham Bell, letter to Mabel Hubbard Bell, September 12, 1875.

131 *"I always thought"*: Mabel Hubbard Bell, letter to Alexander Graham Bell, September 20, 1875.

131 *Aleck fundamentally agreed*: Alexander Graham Bell, letter to Mabel Hubbard Bell, September 26, 1875.

132 *He began to write*: Alexander Graham Bell, letter to Mabel Hubbard Bell, October 5, 1875.
132 *He could feel*: Ibid.
132 *Soon he reveals*: Ibid.
133 *Mabel's response*: Mabel Hubbard Bell, letter to Alexander Graham Bell, October 13, 1875.
133 *Aleck, who would spend*: Alexander Graham Bell, letter to Mabel Hubbard Bell, October 18, 1875. Emphasis in the original.
134 *Beginning on Monday*: Alexander Graham Bell, letter to Sarah Fuller, November 3, 1875.
134 *At the same time*: Ibid.
134 *But by late October*: Gardiner Greene Hubbard, letter to Alexander Graham Bell, October 29, 1875.
134 *By early November*: Alexander Graham Bell, letter to Mabel Hubbard Bell, November 7, 1875.
134 *But by mid-month*: Alexander Graham Bell, letter to Eliza Symonds and Alexander Melville Bell, November 11, 1875.
135 *But tensions had grown*: Eliza Symonds and Alexander Melville Bell, letter to Alexander Graham Bell, November 14, 1875.

PART II

Chapter 8

139 *"I feel like"*: Bruce, *Bell*, 185.
139 *Gray's telegraphic explorations*: Evenson, *Telephone Patent Conspiracy*, 12.
140 *Gray's father died*: Ibid., 13.
140 *That July*: Bruce, *Bell*, 118.
140 *Aleck was preparing*: George Wing, "Physiological Peculiarities of Deafness," *Proceedings of the Eighth Annual Convention of American Instructors of the Deaf and Dumb* (Toronto: Hunter, Rose, 1876), 149; Bruce, *Bell*, 120.
141 *As for Mabel*: Alexander Graham Bell, letter to Gardiner Greene Hubbard, November 23, 1875. Emphasis in the original.
141 *His mother had recently*: Eliza Symonds Bell, letter to Alexander Graham Bell, November 23, 1875.
142 *"—so far away"*: Alexander Graham Bell, letter to Mabel Hubbard Bell, November 25, 1875.
142 *"I am afraid"*: Bruce, *Bell*, 161.

142 *"This day"*: Alexander Graham Bell, letter to Eliza Symonds and Alexander Melville Bell, November 25, 1875.

142 *But as soon as*: Mabel Hubbard Bell, letter to Mary Hatch True, December 2, 1875.

142 *Melville wrote to Mabel*: Bruce, *Bell*, 161.

142 *In the end, even*: Eliza Symonds Bell, letter to Alexander Graham Bell, August 23, 1875.

143 *And she was thrilled*: Eliza Symonds Bell, letter to Alexander Graham, December 12, 1875; Alexander Graham Bell, letter to Eliza Symonds and Alexander Melville Bell, December 7, 1875.

143 *She also requested*: Eliza Symonds Bell, letter to Alexander Graham Bell, December 12, 1875.

143 *In late December*: Alexander Graham Bell, letter to Mabel Hubbard Bell, December 25, 1875.

143 *He made his way*: Ibid.

143 *Toasts rang through*: Alexander Graham Bell, letter to Mabel Hubbard Bell, December 26, 1875.

144 *The next days unfolded*: Ibid.

144 *And he was ready*: Bruce, *Bell*, 163; Watson, *Exploring Life*, 75; Alexander Graham Bell, letter to Mabel Hubbard Bell, January 17, 1876.

144 *Eliza worried*: Eliza Symonds Bell, letter to Alexander Graham Bell, January 23, 1876.

145 *Instead, he gushed*: Alexander Graham Bell, letter to Alexander Melville Bell, January 18, 1876.

145 *As much as he*: Alexander Graham Bell, letter to Mabel Hubbard Bell, January 16, 1876.

145 *"And yet I am"*: Ibid.

145 *Mabel, who was traveling*: Mabel Hubbard Bell, letter to Alexander Graham Bell, January 17, 1876.

145 *He returned, then*: Alexander Graham Bell, letter to Mabel Hubbard Bell, December 28, 1875.

146 *As they drafted*: Watson, *Exploring Life*, 76.

146 *During the day*: Ibid., 80–81.

146 *On January 25*: Seth Shulman, *The Telephone Gambit: Chasing Alexander Graham Bell's Secret* (New York: W. W. Norton, 2008), 74; Bruce, *Bell*, 165.

147 *A few weeks later*: Evenson, *Telephone Patent Conspiracy*, 65–66.

147 *Gardiner, whose entire investment*: Shulman, *The Telephone Gambit*, 74–75. In a later court hearing, Gardiner would claim that Alec gave him permission to do so, but Alec would always claim that it was done without his consent or knowledge.

148 *That both were submitted*: Bruce, *Bell*, 168.
148 *On February 19*: Evenson, *Telephone Patent Conspiracy*, 73.
148 *Alec's lawyers*: Ibid., 42–43.
148 *Soon enough*: Ibid., 88.
148 *But since the cash blotter*: Ibid., 78–79.
149 *As he traveled*: Alexander Graham Bell, letter to Alexander Melville Bell, February 25, 1876.
150 *Alec then edited*: Alexander Graham Bell, letter to Alexander Melville Bell, February 29, 1876; Evenson, *Telephone Patent Conspiracy*, 81.
150 *Alec wrote to*: Alexander Graham Bell, letter to Alexander Melville Bell, February 29, 1876. Emphasis in original.
150 *Back at Exeter Place*: Mackenzie, *Bell*, 82.
150 *The main problem*: Bernard S. Finn, "Alexander Graham Bell's Experiments with the Variable-Resistance Transmitter," *Smithsonian Journal of History* 1, no. 4 (1966): 1–16.
151 *"Mr. Watson, come here"*: Marian (Daisy) Bell, "Comments About First Words Used Over the Telephone," June 13, 1932.
151 *They switched places*: Bruce, *Bell*, 181.
152 *He heard the words*: Ibid.
152 *Now they went for*: Watson, *Exploring Life*, 79; Finn, "Alexander Graham Bell's Experiments," 8–9.
152 *It wasn't enough*: Gardiner Greene Hubbard, letter to Alexander Graham Bell, April 26, 1876.
153 *That night, from home*: Alexander Graham Bell, letter to Mabel Hubbard Bell, May 6, 1876.
153 *"I want to marry"*: Ibid.
153 *"Mr. Bell, are you going"*: Bruce, *Bell*, 187.

Chapter 9

154 *"I shall feel"*: Alexander Graham Bell, letter to Mabel Hubbard Bell, June 18, 1976.
154 *"have the work"*: Alexander Graham Bell, letter to Alexander Melville Bell, October 16, 1875.
155 *"necessary work"*: Alexander M. Bell, letter to Eliza Symonds Bell, April 5, 1876; Bruce, *Bell*, 190.
155 *"the more I examine"*: Alexander Graham Bell, letter to Mabel Hubbard Bell, probably June 1876.
155 *These were the styles*: Watson, *Exploring Life*, 84.

155 *Mabel was on her father's*: Alexander Graham Bell, letter to Eliza Symonds Bell, June 18, 1876.

156 *On the train*: Ibid.; Alexander Graham Bell, letter to Mabel Hubbard Bell, June 18, 1876.

156 *"I love you"*: Ibid.

156 *He arrived Sunday*: Alexander Graham Bell, letter to Mabel Hubbard Bell, June 21, 1876.

156 *As he waited*: Ibid.

156 *And so he called*: Ibid.

157 *Thomson and his wife*: Ibid.

157 *But then he became*: Ibid.

157 *It was then*: Alexander Graham Bell, "The Pre-Commercial Period of the Telephone," 44; Alexander Graham Bell, letter to Mabel Hubbard Bell, June 21, 1876.

157 *Alec hated*: Alexander Graham Bell, letter to Mabel Hubbard Bell, June 21, 1876.

158 *In Cambridge, Mabel*: Mabel Hubbard Bell, letter to Alexander Graham Bell, June 19, 1876.

158 *"I wish my own"*: Mabel Hubbard Bell, letter to Eliza Symonds Bell, May 27, 1876.

158 *And she was attending*: Mabel Hubbard Bell, letter #2 to Alexander Graham Bell, June 19, 1876.

158 *When her excitement overruled*: Mabel Hubbard Bell, letter to Alexander Graham Bell, June 22, 1876.

158 *He worked all day*: Alexander Graham Bell, letter to Mabel Hubbard Bell, June 21, 1876; Alexander Graham Bell, letter to Mabel Hubbard Bell, June 22, 1876.

159 *Gray began*: Alexander Graham Bell, letter to Eliza Symonds and Alexander Melville Bell, June 27, 1876.

159 *They wore full suits*: Bruce, *Bell*, 194–95.

159 *"Mr. Bell"*: Alexander Graham Bell, "The Pre-Commercial Period of the Telephone," 45.

160 *By the time*: Ibid.; Helen Keller, *Midstream: My Later Life* (Garden City, NY: Sun Dial Press, 1937), 116–17; James C. Watson, "The Bell Telephone Company" (circular), February 25, 1879; Alexander Graham Bell, letter to Mabel Hubbard Bell, June 18, 1876.

160 *In the education*: *The Bell Telephone,* 97–100.

161 *He hoped for two things*: Alexander Graham Bell, letter to Eliza Symonds and Alexander Melville Bell, June 27, 1876.

161 *Alec, who was not listening*: Ibid.; Alexander Graham Bell, "The Pre-Commercial Period of the Telephone," 37–57.

161 *Alec was continuing*: Ibid.

161 *Back at the receiving*: Ibid.; Shulman, *The Telephone Gambit*, 191.

161 *Gray thought*: Shulman, *The Telephone Gambit*, 192.

161 *Not even Alec*: Candice Millard, *Destiny of the Republic: A Tale of Madness, Medicine and the Murder of a President* (New York: Doubleday, 2011), 77.

162 *Elisha Gray, a few months*: Evenson, *Telephone Patent Conspiracy*, 119.

162 *No one had yet*: Ibid., 119–23.

163 *"It's nothing more"*: James C. Watson, "The Bell Telephone."

163 *Months later*: Edwin Grosvenor and Morgan Wesson, *Alexander Graham Bell: The Life and Times of the Man Who Invented the Telephone* (New York: Harry Abrams, 1997), 74.

163 *But long before that*: Shulman, *The Telephone Gambit*, 194.

163 *By the fall*: Watson, *Exploring Life*, 86–91.

164 *He was working on*: Bruce, *Bell*, 209.

164 *He was also more in love*: Watson, *Exploring Life*, 108–09.

164 *While Alec quietly succumbed*: Bruce, *Bell*, 209; Alexander Graham Bell, letter to Mabel Hubbard Bell, November 14, 1876.

164 *But despite the rosy ideas*: Caroline A. Yale, *Years of Building: Memories of a Pioneer in a Special Field of Education* (New York: Dial Press, 1931), 58.

165 *Not even Alec*: Ibid., 58.

165 *It was a short distance*: Watson, "Notes on the Early Days," 340; Watson, *Exploring Life*, 93–96.

166 *Thomas was dutiful*: Watson, *Exploring Life*, 100.

166 *Remembering his childhood*: Ibid.

167 *Instead, he went*: Ibid.

167 *With permission from*: Bruce, *Bell*, 205; Alexander Graham Bell, "The Pre-Commercial Period of the Telephone," 37–57. The story in "The Pre-Commercial Period . . ." is slightly different. It says they borrowed a preexisting line from the Observatory and used it at night when it wasn't needed for telegraph signals.

167 *The* Boston Post: Bruce, *Bell*, 207.

168 *Publicity was picking up*: Bruce, *Bell*, 208; Gardiner Greene Hubbard, letter to Thomas Watson, January 20, 1877.

168 *For once, Alec listened*: Watson, *Exploring Life*, 109.

168 *On February 12*: Mabel Hubbard Bell, letter to Gertrude Hubbard, undated.

168 *By now, Alec*: Bruce, *Bell*, 219.

169 *He began by explaining*: Mabel Hubbard Bell, letter to Gertrude Hubbard, undated.

169 *He transmitted*: Gray (*Reluctant Genius*, 157) makes it seem like Watson sang these, and he did sing these songs frequently, but newspaper records note them as from an organ. No source from Gray.

169 *Before he started*: Gray, *Reluctant Genius*, 157.

169 *From the box came*: Bruce, *Bell*, 217.

169 *For a moment, Mabel*: Mabel Hubbard Bell, letter to Gertrude Hubbard, undated.

170 *They hushed again*: Bruce, *Bell*, 217.

170 *"I wasn't listening"*: "Sent by Telephone," *Boston Globe*, February 12, 1877.

170 *In an editorial*: Ibid.

170 *In general, people*: Watson, *Exploring Life*, 98.

171 *Thomas saw the gift*: Ibid., 111.

171 *Cousin Mary weighed in*: Mabel Hubbard Bell, letter to Gertrude Hubbard, February 1877.

171 *But Mabel kept*: Bruce, *Bell*, 219.

171 *After the Salem lectures*: Watson, *Exploring Life*, 113–14.

171 *The duo's demonstrations*: Gray, *Reluctant Genius*, 156.

172 *"I went to his office"*: Bruce, *Bell*, 228. Emphasis in original.

Chapter 10

173 *"Time and distance"*: Grosvenor, *Alexander Graham Bell*, 86.

173 *On July 9, 1877*: Bruce, *Bell*, 235.

173 *There were an estimated*: Grosvenor, *Alexander Graham Bell*, 83.

173 *On July 11*: Bruce, *Bell*, 233.

174 *Now Alec and Mabel*: Mabel Hubbard Bell, letter to Gertrude Hubbard, July 20, 1877.

174 *Mabel thought her*: Bruce, *Bell*, 234; Toward, *Mabel Bell*, 48; Marian (Daisy) Bell, "Notes on Alexander Graham Bell."

174 *Alec's family had planned*: Mabel Hubbard Bell, letter to Gertrude Hubbard, July 23, 1877; E. Pauline Johnson, "The Story of the First Telephone," *Beinn Bhreagh Recorder* 7, 440. Alexander Graham Bell Family Papers, Manuscript Division, Library of Congress.

174 *And so Chief Johnson*: According to Johnson ("The Story of the First Telephone"), Chief Johnson said, "*Saco gatchi; ska na ka?*" or "Good greeting, cousin; how are you?" (Johnson's translation).

174 *Mabel marveled*: Mabel Hubbard Bell, letter to Gertrude Hubbard, July 23, 1877.

175 *In early August*: Grosvenor, *Alexander Graham Bell*, 88; Bruce, *Bell*, 235.

175 *On board, there were*: Mabel Hubbard Bell, letter to Eliza Symonds Bell, August 14, 1877.
175 *They landed in time*: Grosvenor, *Alexander Graham Bell*, 88; Bruce, *Bell*, 242–43.
175 *In October*: Tom Standage, *The Victorian Internet: The Remarkable Story of the Telegraph and the Nineteenth Century's On-Line Pioneers* (New York: Walker, 1998), 199.
175 *Here they planned*: Mabel Hubbard Bell, letter to Eliza Symonds Bell, October 10, 1877.
175 *"Alec appears to think"*: Bruce, *Bell*, 236.
175 *But never mind*: Mabel Hubbard Bell, letter to Gertrude Hubbard, October 1, 1877.
175 *When it became clear*: Mabel Hubbard Bell, letter to Eliza Symonds Bell, October 10, 1877.
176 *They went every day*: Bruce, *Bell*, 240.
176 *"O Mamma darling"*: Mabel Hubbard Bell, letter to Gertrude Hubbard, August 23, 1877.
176 *Climbing back over*: Mabel Hubbard Bell, letter to Eliza Symonds Bell, October 10, 1877.
177 *He went back to*: Mabel Hubbard Bell, letter to Gertrude Hubbard, October 1, 1877.
177 *Mabel worked on*: Mabel Hubbard Bell, letter to Eliza Symonds Bell, October 10, 1877.
177 *Then, turning to*: Mabel Hubbard Bell, letter to Gertrude Hubbard, October 8, 1877.
178 *After their honeymoon*: Grosvenor, *Alexander Graham Bell*, 90; Bruce, *Bell*, 240–44.
178 *As the pair settled*: Toward, *Mabel Bell*, 51.
178 *Despite all the effort*: Toward, *Mabel Bell*, 51; Mabel Hubbard Bell, letter to Gertrude Hubbard, August 15, 1877.
178 *But Alec's star*: Bruce, *Bell*, 242.
178 *But that's not to say*: Mabel Hubbard Bell, letter to Gertrude Hubbard, January 9, 1878; Gray, *Reluctant Genius*, 181.
179 *Mabel's dress arrived*: Toward, *Mabel Bell*, 60.
179 *Meanwhile, on the evening*: Bruce, *Bell*, 239.
179 *He waited*: Bruce, *Bell*, 241–44.
179 *And then, after Alec*: Ibid.
180 *They listened to*: Ibid.
180 *Queen Victoria called*: Ibid.

180 *Alec and Mabel settled*: Mabel Hubbard Bell, diary, March, 14, 1880; Mabel Hubbard Bell, letter to Eliza Symonds Bell, May 24, 1878.
180 *Alec took to his*: Bruce, *Bell*, 237; Alexander Graham Bell, letter to Mabel Hubbard Bell, September 5, 1878.
180 *Outside the walls*: Bruce, *Bell*, 242.
181 *In America, it was satirized*: Grosvenor, *Alexander Graham Bell*, 84.

Chapter 11

182 *"I do not see"*: Alexander Graham Bell, letter to Mabel Hubbard Bell, September 9, 1878.
182 *In Great Britain*: Gertrude Hubbard, letter to Eliza Symonds Bell, April 17, 1878.
182 *"likely to inaugurate"*: Bruce, *Bell*, 256.
182 *As it was*: *Greenock Advertiser*, May 21, 1878; *Greenock Telegraph*, May 21, 1878.
183 *When Alec imagined*: Alexander Graham Bell, letter to Mabel Hubbard Bell, September 5, 1878.
183 *Alec's fears proved*: Ibid.
186 *When he opened*: Richard Winefield, *Never the Twain Shall Meet: Bell, Gallaudet, and the Communications Debate* (Washington, DC: Gallaudet University Press, 1987), 26.
186 *At an oralist school*: Edward Miner Gallaudet, "Report of the President on the Systems of Deaf-Mute Instruction Pursued in Europe," *Tenth Annual Report of the Columbia Institution* (Washington, DC: 1867), 19.
186 *But it wasn't just*: Ibid., 23.
187 *Gallaudet began to see*: Ibid., 52; Edward Miner Gallaudet, "Results of Articulation Teaching at Northampton," *American Annals of the Deaf and Dumb* 19, no. 3 (July 1874): 138.
187 *Back in 1874*: Gallaudet, "Results of Articulation Teaching at Northampton," 138.
187 *He would come to*: Ibid., 141.
188 *Gallaudet came to*: Gallaudet, "Report of the President," 53.
189 *Today, this is understood*: Sanjay Gulati, "Language Deprivation Syndrome," Brown University, April 1, 2014. "If you don't have an L1, you have brain damage. It cannot be resolved."
189 *In 1878, as Alec*: Charles Strong Perry, "The Acquisition of Language," *American Annals of the Deaf and Dumb* 22, no. 2 (April 1877): 72.
190 *For the first time*: Alexander Graham Bell, letter to Mabel Hubbard Bell, September 5, 1878.

190 *Mabel had stayed*: Ibid.

191 *She didn't want to*: Toward, *Mabel Bell*, 236–37.

191 *Mabel wanted nothing*: Toward, *Mabel Bell*, 236–37. Later she wrote, "Above all things I was antagonistic to my husband's efforts to keep up his association with the deaf and to continue his teaching of them."

191 *Not that it mattered*: Alexander Graham Bell, letter to Mabel Hubbard Bell, September 5, 1878.

191 *Mabel wrote that*: Alexander Graham Bell, letter to Mabel Hubbard Bell, September 9, 1878.

191 *Alec snapped back*: Ibid.

192 *His response was*: Ibid.

192 *Alec had wanted*: Gardiner Greene Hubbard, letter to Mabel Hubbard Bell, November 2, 1878; Alexander Graham Bell, letter to Mabel Hubbard Bell, September 5, 1878.

192 *Alec and Mabel were*: Alexander Graham Bell, letter to Mabel Hubbard Bell, September 9, 1878.

Chapter 12

194 *"They have hacked"*: Beauchamp, *Invented by Law*, 65.

194 *Alec hadn't even*: Ibid., 52.

194 *The American Speaking Telephone*: Bruce, *Bell*, 262.; Gardiner Greene Hubbard, letter to Alexander Graham Bell, March 13, 1878.

195 *Not only that*: Grosvenor, *Alexander Graham Bell*, 92.

195 *Soon, the American Speaking*: Gray, *Reluctant Genius*, 193; Evenson, *Telephone Patent Conspiracy*, 129; Beauchamp, *Invented by Law,* 54–55.

195 *Those who had poured*: Gray, *Reluctant Genius*, 189; Evenson, *Telephone Patent Conspiracy*, 124; Beauchamp, *Invented by Law*, 52; Grosvenor, *Alexander Graham Bell*, 92; Evenson, *Telephone Patent Conspiracy*, 124.

196 *He believed that*: Bruce, *Bell*, 264–66. He didn't harp on the metallic diaphragm, because he didn't think this was as important of an innovation as the permanent magnet.

196 *With Western Union's*: Bruce, *Bell*, 267.

196 *All Alec wanted*: Alexander Graham Bell, letter to Mabel Hubbard Bell, September 9, 1878.

197 *On the shore*: Watson, *Exploring Life*, 152; Gray, *Reluctant Genius*, 190.

197 *"I went with him"*: Gray, *Reluctant Genius*, 190.

197 *Days later, still*: Alexander Graham Bell, letter to Mabel Hubbard Bell, November 14, 1878.

198 *That same night*: *Complimentary Reception and Banquet to Elisha Gray, PhD, Inventor of the Telephone, at Highland Park, November 15, 1878* (Chicago: C. E. Southard, 1879). Smithsonian Museum of American History.
198 *At the banquet*: Ibid.
198 *One of the arrangers*: Ibid.
198 *Gray listened on*: Ibid.
198 *The speech went on*: Ibid.
199 *The day after*: Gray, *Reluctant Genius*, 192.
199 *Though Gardiner and Alec*: Mabel Hubbard Bell, letter to Gertrude Hubbard, November 21, 1878.
199 *Now, with Mabel*: Mabel Hubbard Bell, letter to Eliza Symonds Bell, November 22, 1878.
199 *"The worst is over"*: Ibid.
199 "*my right to see*": Alexander Graham Bell, letter to Alexander Melville Bell, February 29, 1876. Bell does write about the caveat process in an earlier letter (Alexander Graham Bell to Alexander Melville and Eliza Symonds Bell, October 23, 1874), in which he is under the impression that a caveat will remain secret, but it still may be true that he believed he could view someone's if it was in conflict with his patent.
200 *Alec couldn't get*: Alexander Graham Bell, letter to Eliza Symonds Bell, November 23, 1878.
200 *He had seen*: Mabel Hubbard Bell, letter to Eliza Symonds Bell, November 21, 1878.
200 *If the telephone collapsed*: Alexander Graham and Mabel Hubbard Bell, letter to Gertrude Hubbard, November 23, 1878.
200 *"It is too wintery"*: Alexander Graham Bell, letter to Mabel Hubbard Bell, January 21, 1879; Evenson, *Telephone Patent Conspiracy*, 131–32.
200 *"They have hacked"*: Beauchamp, *Invented by Law*, 65.
201 *The company's other lawyer*: Ibid.; Bruce, *Bell*, 267–68.
201 *There was no more talk*: Bruce, *Bell*, 281–82.
201 *On the afternoon of*: Evenson, *Telephone Patent Conspiracy*, 132. Examiner Wilber had waived the requirement.
201 *At the time*: Beauchamp, *Invented by Law*, 58–63.
201 *These two notions*: Ibid.
202 *If inventions were*: Ibid.
202 *"The instruments* talk": Alexander Graham Bell, letter to Mabel Hubbard Bell, January 23, 1879.
203 *"the method of"*: Evenson, *Telephone Patent Conspiracy*, 232.
203 *"They have subjected"*: Beauchamp, *Invented by Law*, 65.

204 *Smith asked Renwick*: Alexander Graham Bell, letter to Mabel Hubbard Bell, January 26, 1879.

204 *"I never saw men"*: Ibid.; Beauchamp, *Invented by Law*, 66.

205 *Alec split his time*: Toward, *Mabel Bell*, 71.

205 *As Alec traveled*: Gray, *Reluctant Genius*, 194.

205 *"Don't let us consent"*: Alexander Graham Bell, letter to Mabel Hubbard Bell, March 14, 1879.

206 *"for I feel sure"*: Bruce, *Bell*, 269. Quoted differently in Gray, *Reluctant Genius*, 196.

206 *"I should explain it"*: *Bell Telephone*, 165–69.

206 *"So far as I know"*: Elisha Gray, letter to Alexander Graham Bell, February 24, 1877.

207 *"For of course"*: Alexander Graham Bell, letter to Elisha Gray, March 2, 1877.

207 *Gray's work toward*: Grosvenor, *Alexander Graham Bell*, 93; Bruce, *Bell*, 269. Emphasis in the original.

207 *In court, the words*: Alexander Graham Bell, letter to Mabel Hubbard Bell, April 8, 1879.

207 *"I'll swear to it"*: Bruce, *Bell*, 269. Quoted differently in Grosvenor, *Alexander Graham Bell*, 93.

208 *Alec gave his own*: Bruce, *Bell*, 270.

208 *In the meantime*: Ibid., 292; Mabel Hubbard Bell, letter to Alexander Graham Bell, March 9, 1879. Emphasis in original.

208 *By summer, Mabel*: Grosvenor, *Alexander Graham Bell*, 97–98.

209 *In the end, Western Union*: Evenson, *Telephone Patent Conspiracy*, 136–37.

209 *seventeen years*: Bruce, *Bell*, 270. Quoted differently—"for the lifetime of the patents"—in Beauchamp, *Invented by Law*, 56.

209 *Gardiner wrote to*: Gardiner Greene Hubbard, letter to Alexander Graham Bell, November 2, 1879.

209 *The stock prices*: Gray, *Reluctant Genius*, 203.

209 *Thomas Watson wrote*: Ibid., 197–98.

209 *over thirty thousand*: Standage, *Victorian Internet*, 199.

209 *On February 15*: Mabel Hubbard Bell, diary. Mabel's diary of the time says "Marianne," though later she is referred to as Marian.

Chapter 13

211 *"Nothing will ever"*: Millard, *Destiny of the Republic*, 258; Alexander Graham Bell, letter to Mabel Hubbard Bell, December 12, 1885. Emphasis in the original.

211 *In Cambridge, the sun*: Mabel Hubbard Bell, letter to Gertrude Hubbard, June 14, 1881.

211 *Mabel found it*: Toward, *Mabel Bell*, 85; Mabel Hubbard Bell, letter to Gertrude Hubbard, June 1881.

211 *Mabel's own role*: Toward, *Mabel Bell*, 85; Mabel Hubbard Bell, letter to Gertrude Hubbard, July 5, 1881.

211 *Elsie was*: Mabel Hubbard Bell, letter to Gertrude Hubbard, June 23, 1881.

211 *Daisy was*: Mabel Hubbard Bell, letter to Gertrude Hubbard, June 28, 1881.

211 *Alec began to*: Mabel Hubbard Bell, letter to Gertrude Hubbard, June 14, 1881.

212 *When Elsie had a party*: Mabel Hubbard Bell, letter to Gertrude Hubbard, June 28, 1881.

212 *"Alec says he would"*: Millard, *Destiny of the Republic*, 219; Mabel Hubbard Bell, letter to Gertrude Hubbard, June 23, 1881.

212 *As Mabel and Alec*: Eliza Symonds Bell, letter to Alexander Graham Bell, July 8, 1881; Mabel Hubbard Bell, letter to Gertrude Hubbard, June 23, 1881.

212 *But on July 3*: "A Dark Deed," *Boston Globe*, morning edition, July 3, 1881.

213 *He wasn't the only*: Millard, *Destiny of the Republic*, 177.

213 *His assistant, Charles*: Ibid., 176.

213 *"I cannot possibly"*: Mabel Hubbard Bell, letter to Gertrude Hubbard, July 8, 1881.

213 *Then he read*: Bruce, *Bell*, 344–45.

214 *On July 14*: Mackenzie, *Bell*, 235.

214 *Inside, Alec met*: Alexander Graham Bell, letter to Mabel Hubbard Bell, July 17, 1881.

214 *Alec and Charles*: Eliza Symonds Bell, letter to Alexander Graham Bell, July 8, 1881.

214 *Alec would hear*: Jeffrey S. Reznick and Lenore Barbian, "'The President Is Somewhat Restless . . .': Enter Bell," US National Library of Medicine, July 26, 2013, https://circulatingnow.nlm.nih.gov/2013/07/26/the-president-is-somewhat-restless-enter-bell/.

214 *When Alec reached back*: Alexander Graham Bell, letter to Mabel Hubbard Bell, July 17, 1881.

215 *Before each test*: Mackenzie, *Bell*, 236.

215 *That day, Mabel*: Alexander Graham Bell, letter to Mabel Hubbard Bell, July 16, 1881.

215 *Alec worried about*: Alexander Graham Bell, letter to Mabel Hubbard Bell, July 17, 1881.

215 *Mabel wrote back*: Mabel Hubbard Bell, letter to Eliza Symonds Bell, July 17, 1881.

215 *"I fear she may"*: Toward, *Mabel Bell*, 86; Mabel Hubbard Bell, letter to Gertrude Hubbard, July 5, 1881.

216 *Soon, Alec could*: Millard, *Destiny of the Republic*, 218.

216 *That night, Mabel*: Mabel Hubbard Bell, letter to Alexander Graham Bell, July 22, 1881.

216 *But earlier that day*: Millard, *Destiny of the Republic*, 226.

216 *The next day*: Ibid., 230.

217 *By July 24*: Mabel Hubbard Bell, letter to Alexander Graham Bell, July 24, 1881.

217 *On the twenty-fifth*: Alexander Graham Bell, letter to Mabel Hubbard Bell, July 26, 1881.

217 *Mabel hoped*: Mabel Hubbard Bell, letter to Alexander Graham Bell, July 25, 1881.

217 *Alec slept until*: Alexander Graham Bell, letter to Mabel Hubbard Bell, July 26, 1881.

218 *Alec brought his*: Ibid.

218 *Now that look*: Ibid.; Bruce, *Bell*, 346.

218 *Bliss operated*: Mackenzie, *Bell*, 237. Mackenzie says the president's eyes never left Alec, but in a letter, Alec says he was seated behind the president. These aren't necessarily incompatible but might suggest an inconsistency.

219 *In Cambridge, Mabel*: Mabel Hubbard Bell, letter to Alexander Graham Bell, July 29, 1881.

219 *It rained every day*: Mabel Hubbard Bell, letter to Eliza Symonds Bell, July 29, 1881; Mabel Hubbard Bell, letter to Alexander Graham Bell, July 29, 1881.

220 *Back at the Volta Lab*: Millard, *Destiny of the Republic*, 258; Bruce, *Bell*, 347.

220 *"I don't want"*: Millard, *Destiny of the Republic,* 260.

220 *The next day*: Ibid., 265.

220 *On the night of*: Bruce, *Bell*, 347.

220 *"I hope indeed"*: Ibid.

221 *In the end*: Mabel Hubbard Bell, diary, May 20, 1884.

221 *Daisy's first memory*: Marian (Daisy) Bell, "My First Impressions about My Father and Mother," February 6, 1927.

221 *"We smile at"*: Alexander Graham Bell, letter to Mabel Hubbard Bell, September 19, 1881.

222 *He wrote to her*: Alexander Graham Bell, letter to Mabel Hubbard Bell, September 24, 1881.

222 *Elsie, in particular*: Ibid. Emphasis in the original.

222 *"For my own part"*: Ibid.
222 *"You know Alec"*: Alexander Graham Bell and Mabel Hubbard, letter to Eliza Symonds Bell, September 1881.
222 *She took breaks*: Ibid.
222 *Alec made his own*: Ibid.
223 *By late October*: Alexander Graham Bell, letter to Mabel Hubbard Bell, October 29, 1881.
223 *Alec loved animals*: Marian (Daisy) Bell, memories of Mabel Hubbard Bell, July 3, 1928.
223 *Small lessons*: Mabel Hubbard Bell, letter to Alexander Graham Bell, June 22, 1883.
223 *He thought the children*: Alexander Graham Bell, letter to Mabel Hubbard Bell, September 4, 1883.
224 *He hoped the children*: Ibid.

PART III

Chapter 14

227 *"A new epoch"*: E. A. Hodgson, *The Deaf-Mutes' Journal* IX, no. 35 (August 26, 1880).
228 *The Abbé Tarra*: James Denison, "Impressions of the Milan Convention," *American Annals of the Deaf and Dumb* 26, no. 1 (January 1881): 47.
228 *At the outset*: Richard G. Brill, *International Congresses on Education of the Deaf: An Analytical History, 1878–1980* (Washington, DC: Gallaudet College Press, 1984), 19.
229 *Before that moment*: *Speech for the Deaf: Essays Written for the Milan International Congress: Proceedings and Resolutions* (London: W. H. Allen, 1880), 72.
229 *"A 'Combined' system"*: Ibid., 76.
230 *To explain herself*: Ibid., 80.
230 *Instead, her elementary*: Ibid., 81.
230 *Actual learning was delayed*: Ibid., 82.
230 *In the days following*: Richard Elliot, "The Milan Congress and the Future of the Education of the Deaf and Dumb," *American Annals of the Deaf and Dumb* 27, no. 3 (July 1882): 152.
231 *The advanced students*: Ibid.
231 *"excited the admiration"*: Denison, "Impressions of the Milan Convention," 47.
231 *And while certain*: Ibid., 44–45.

232 *Gallaudet soon abandoned*: Brill, *International Congresses*, 23; Edward Miner Gallaudet, "Rejoinder to Padre Marchio," *American Annals of the Deaf and Dumb* 26, no. 3 (1881): 164.

232 *He wasn't the only one*: Harlan Lane, *When the Mind Hears: A History of the Deaf* (New York: Random House, 1984), 387.

232 *At the end*: Denison, "Impressions of the Milan Convention," 47; Brill, *International Congresses*, 20.

232 *As the attendees returned*: Brill, *International Congresses*, 18.

233 *But the Milan Conference*: Jack R. Gannon, *Deaf Heritage: A Narrative of Deaf America* (Silver Spring, MD: National Association of the Deaf, 1981), 59–63.

233 *It wasn't without conflict*: Hodgson, *Deaf-Mutes' Journal*, 2.

233 *On August 24, 1880*: "The National Convention," *Deaf-Mutes' Journal* IX, no. 36 (September 2, 1880): 3; A Young Member, "Convention Jottings," *Deaf-Mutes' Journal* IX, no. 37 (September 9, 1880): 3.

233 *They knew each other's names*: "Convention Jottings," *Deaf-Mutes' Journal*.

233 *Later stories would summarize*: "Columbus," *Deaf-Mutes' Journal* IX, no. 36 (September 2, 1880): 3.

234 *As those with stature*: "Convention Personalities, Called by 'We, Us & Co.,'" *Deaf-Mutes' Journal* IX, no. 37 (September 9, 1880): 3.

234 *One group*: "Columbus," *Deaf-Mutes' Journal*, September 2, 1880, 3.

234 *"[We] were astonished"*: Ibid.

234 *At the conference*: "Columbus," *Deaf-Mutes' Journal*.

234 *On the first day*: Ibid.

235 "Fellow graduates": "The National Convention," *Deaf-Mutes' Journal*.

235 *Wrote one regular columnist*: "Convention Personalities," *Deaf-Mutes' Journal*.

236 *Alec had also turned*: Alexander Graham Bell, letter to Edward Miner Gallaudet, October 11, 1880.

237 *Instead, commentary was provided*: Mr. Why, "Something for Teachers of Articulation to Consider," *Deaf-Mutes' Journal* X, no. 3 (January 20, 1881): 1.

237 *"What must be thought"*: Ibid.

237 *Pointing out that most*: Ibid.

238 *Denison had observed*: James Denison, letter to Alexander Graham Bell, November 22, 1882.

238 *"It seems to me"*: Ibid.

238 *"How important as it"*: Ibid.

238 *"He has proved himself"*: Ibid.

239 *"We need your words"*: Sarah Fuller, letter to Alexander Graham Bell, February 2, 1884.

240 *"the result of this"*: John Hitz, *Dr. A. Graham Bell's Private Experimental School* (Washington, DC: Sanders Printing Office, 1898), 9.

240 *But the bigger tragedy*: Alexander Graham Bell, speech, Fifth National Conference of Principals and Superintendents of Institutions for Deaf-Mutes, Faribault, MN, July 9–13, 1884.

241 *Alec worked only with*: Alexander Graham Bell, notebook, 1883, 8–9.

241 *Day in and day out*: Alexander Graham Bell, letter to Mabel Hubbard Bell, September 4, 1883.

241 *In early October*: Alexander Graham Bell, notebook, 1883, 9.

242 *Later that day*: Ibid., 9–12.

242 *Daisy and Elsie*: Ibid., 12.

242 *But already there were*: Mabel Hubbard Bell, diary, September 16 (likely 1883, possibly 1884).

Chapter 15

244 *"[The students] not only have"*: Alexander Graham Bell, letter to Mrs. Bingham, November 8, 1885.

244 *In September 1883*: Mabel Hubbard Bell, diary, September 9, 1883.

244 *Back when he began*: Bruce, *Bell*, 409. This effort began in 1878.

244 *The phenomenon of heredity*: Brian H. Greenwald and John Vickery Van Cleve, "'A Deaf Variety of the Human Race': Historical Memory, Alexander Graham Bell, and Eugenics," *Journal of the Gilded Age and Progressive Era* 14, no. 1 (2015): 32–33.

245 *What did surprise Alec*: Alexander Graham Bell, *Memoir upon the Formation of a Deaf Variety of the Human Race* (n.p.: Alexander Graham Bell Association for the Deaf, 1969), 40. Emphasis in original.

245 *By the early 1880s*: Douglas C. Baynton, *Defectives in the Land: Disability and Immigration in the Age of Eugenics* (Chicago: University of Chicago Press, 2016), 12.

245 *But Darwin took*: All Darwin quotes: Charles Darwin, *The Descent of Man, and Selection in Relation to Sex* (New York: D. Appleton, 1871), 161–62.

246 *When Alec first*: Alexander Graham Bell, letter to Edward Miner Gallaudet, October 11, 1880. Some of the punctuation (excess dashes) of this quote has been omitted.

247 *In late October*: Mabel Hubbard Bell, diary, October 21, 1883.

247 *That night, after*: Ibid.

247 *Before Mabel joined*: Ibid.

248 *In New Haven*: "Scientists in Council: Topics of Discussion Before the

National Academy of Sciences," *New York Times*, November 14, 1883; Alexander Graham Bell, *Memoir*, 3; "Papers on Sciences," *New York Tribune*, November 14, 1883. The number in the audience is recorded as almost one hundred, not including students and postgraduates. Over one hundred is my inference.

248 *He said that*: Ibid.

249 *The night of November 18*: Mabel Hubbard Bell, diary, February 3, 1884.

249 "*It was so pretty*": Ibid.; Gray, *Reluctant Genius*, 232.

249 *Alec arrived home*: Gray, *Reluctant Genius,* 232.

249 *In the wake of*: "College Chronicle. Early Ideas about Deaf-Mutes. Ghosts and Gas." *Deaf-Mutes' Journal* XIII, no. 6 (February 7, 1884): 2. It was summarized: "In many countries, they were destroyed; in France, the birth of a deaf child was considered as a great disgrace, a dire infliction from heaven on the parents; the Code of Justinian deprived them of all civil rights, and even large-hearted St. Augustine held that the congenital mute had no claim to any place in paradise, and passes the severe judgment that 'deafness from birth makes faith impossible, since he who is born deaf can neither hear the Word, nor learn how to read it.'"

250 *But this writer*: Ibid.

250 *The narrative wrests*: Ibid.

251 *And they were pooling*: Brian H. Greenwald and John Vickery Van Cleve, eds., *A Fair Chance in the Race of Life: The Role of Gallaudet University in Deaf History* (Washington, DC: Gallaudet University Press, 2008), 33.

251 *Alec kept pushing*: Greenwald and Van Cleve, *Fair Chance*, 33.

251 *"We believe it has"*: E. A. Hodgson, "Prof. Bell's Lecture," *Deaf-Mutes' Journal* XII, no. 46 (November 15, 1883): 2.

252 *On the same day*: "In Reference to Deaf Mutes," *Evening Star* (Washington, DC), December 31, 1884, 2.

252 *An article soon ran*: J. F. J. Fresch, "Perpetuating Deaf Mutes: Prof Bell's Theory Energetically Disputed," *Washington Post*, January 4, 1885.

252 *If Alec was following*: Alexander Graham Bell, letter to Mabel Hubbard Bell, December 12, 1885.

252 *Alec believed his own*: Ibid.

253 *Alec loved his school*: Mabel Hubbard Bell, journal, 1885.

253 *The second year*: Alexander Graham Bell, notebook for Bell's school; Record Book of Mr. Bell's, 1884–1885, 49.

253 *The deaf students gathered*: Alexander Graham Bell, notebook for Bell's school.

254 *At the school*: Ibid.

254 *Alec taught them*: Alexander Graham Bell, Record Book of Mr. Bell's Private School, 1884–85, 30–43.

254 *In the summer*: Mackenzie, *Bell*, 256–57. They were shipwrecked on the second leg of their journey, and never did make it to Newfoundland, returning instead to Cape Breton.

254 *But as they relished*: Beauchamp, *Invented by Law*, 88; Harvey Gresham Hudspeth, "'One Percent Inspiration and 99 Percent Tracing Paper': The Pan-Electric Scandal and the Making of a Circuit Court Judge, April–November 1886," *Essays in Economic and Business History* 23, no. 1 (2005); Evenson, *Telephone Patent Conspiracy*, 191.

256 *His full articulation*: Alexander Graham Bell, letter to Mrs. Bingham, November 8, 1885 (copy). It's likely that Mrs. Bingham was the mother of one of the students in the class, Floyd Bingham, but this hasn't been confirmed.

256 *It was the same*: Beauchamp, *Invented by Law,* 89–90. Officially, the reason was that "the standard practice of referral to the Interior Department had not been followed."

257 *"[The students] not only have"*: Alexander Graham Bell, letter to Mrs. Bingham, November 8, 1885.

258 *"from the particular"*: Alexander Graham Bell, "Fallacies Concerning the Deaf," *American Annals of the Deaf and Dumb* 29, no. 1 (January, 1884): 61.

258 *At 2:00 p.m. on Wednesday*: Bruce, *Bell*, 390.

259 *"His little school"*: Mabel Hubbard Bell, diary, November 19, 1885.

259 *In this moment*: Ibid.

Chapter 16

260 *"There is a pleasure"*: Marian (Daisy) Bell, memories of Alexander Graham Bell, July 4, 1926.

260 *Three years earlier*: Helen Keller, *Story of My Life* (New York: W. W. Norton, 2004), 13–16; Herrmann, *Helen Keller,* 9; "brain fever" was a sort of Victorian-era catchall.

260 *But then her health*: Herrmann, *Helen Keller,* 9–11.

261 *He traveled simply*: Alexander Graham Bell, letter to Mabel Hubbard Bell, November 29, 1885.

261 *His letters to Mabel*: Mabel Hubbard Bell, letter to Alexander Graham Bell, November 30, 1885.

261 *The island moved*: Alexander Graham Bell, letter to Mabel Hubbard Bell, December 3, 1885.

261 *Many in New England's*: Nora Ellen Groce, *Everyone Here Spoke Sign Language: Hereditary Deafness on Martha's Vineyard* (Cambridge, MA: Harvard University Press, 1985), 3.

261 *In the town of*: Greenwald and Van Cleve, "'A Deaf Variety of the Human Race,'" 34.

261 *The island, which*: Groce, *Everyone Here*, 8–9.

261 *The books that remained*: Alexander Graham Bell, letter to Mabel Hubbard Bell, December 8, 1885.

261 *As he was copying*: Groce, *Everyone Here*, 4–5.

262 *Within four days*: Alexander Graham Bell, letter to Mabel Hubbard Bell, December 8, 1885. Emphasis in original.

263 *He worked compulsively*: Ibid.

263 *Alec took up quiet*: Ibid.

263 *From outside the Sailors'*: Ibid.

263 *He thought of Mabel*: Alexander Graham Bell, letter to Mabel Hubbard Bell, December 12, 1885.

264 *In Alabama, Helen*: Keller, *Story of My Life*, 17. "I had noticed that my mother and my friends did not use signs as I did when they wanted anything done, but talked with their mouths. Sometimes I stood between two persons who were conversing and touched their lips. I could not understand, and was vexed. I moved my lips and gesticulated frantically without result. This made me so angry at times that I kicked and screamed until I was exhausted."

264 *In the summer of 1886*: This date is often noted as 1887 (Bruce; Herrmann), but Bell's note to Gallaudet is dated 1886. Joseph P. Lash claims that it is over the summer; in *Story of My Life* (p. 23), Keller says she is about six years old.

265 *And another inventor*: Bruce, *Bell*, 271–72.

265 *Meanwhile, several older*: Beauchamp, *Invented by Law*, 78–79.

266 *Alec received correspondence*: E. S. Doolittle, letter to Alexander Graham Bell, September 24, 1883.

266 *They worried their child*: A. I. Hipkins, letter to Alexander Graham Bell, November 21, 1883.

266 *The youngest deaf writers*: Lottie F. Bailey, letter to Alexander Graham Bell, September 25, 1883.

266 *He heard, too, from*: Albert H. Hutchins, letter to Alexander Graham Bell, September 25, 1883.

266 *Alec asked the National Bell*: Bruce, *Bell*, 399.

266 *He gave money*: Sarah Fuller, letter to Alexander Graham Bell, November 3, 1885; Alexander Graham Bell, letter to Sarah Fuller, June 2, 1888.

266 *When one deaf man*: Bruce, *Bell*, 296.

267 *Alec wrote publicly*: Alexander Graham Bell, "New View of Miss Pancoast's Case: Deaf Mutes Do Not Think in English—Their Language Is as Distinct as German or French," *New York Tribune*, March 16, 1886, 4.

267 *He understood her*: Richard McGann, personal interview, February 27, 2020, secondhand memory told to Robert Smithdas by Helen Keller.

267 *She loved him*: She says that he understood her signs, but since she didn't have any apparent exposure to Sign at this point, I'm guessing what he actually understood was her gestures or possibly her home signs.

267 *Later, she wrote*: Keller, *Story of My Life*, 24.

267 *Alec invited Helen*: Alexander Graham Bell, letter to E. M. Gallaudet, *Letters from Famous Americans*, Helen Keller Archives, American Foundation for the Blind. This letter is archived as being written to Thomas Gallaudet, but he was long dead.

267 *Alec advised Mr. Keller*: Keller, *Story of My Life*, 24.

269 *When the word was*: Ibid., 23.

269 *And then, the words*: Ibid., 24.

269 *"Dear Mr. Bell"*: Helen Keller, letter to Alexander Graham Bell, November 1887.

Chapter 17

271 *"Alec talked genealogy"*: Gray, *Reluctant Genius*, 231.

272 *When Alec befriended*: Ballin, *The Deaf Mute Howls*, 41.

272 *When they traveled*: Toward, *Mabel Bell*, 126.

272 *Mabel felt that she*: Bruce, *Bell*, 323; Mabel Hubbard Bell, letter to Alexander Graham Bell, December 3, 1889.

272 *As Alec became more*: Bruce, *Bell*, 308; Mabel Hubbard Bell, letter to Alexander Graham Bell, July 9, 1895.

273 *Alec may have believed*: Bruce, *Bell*, 309; Alexander Graham Bell, letter to Mabel Hubbard Bell, May 5, 1890.

273 *When he was absorbed*: Alexander Graham Bell, letter to Mabel Hubbard Bell, May 23, 1887.

273 *In 1888*: Bruce, *Bell*, 423.

273 *He also became*: Beauchamp, *Invented by Law*, 99.

274 *As Alec was pursuing*: Greenwald and Van Cleve, "'A Deaf Variety of the Human Race,'" 34–35.

274 *At the 1889 convention*: *Proceedings of the Third Convention of the National Association of the Deaf, held at the National Deaf-Mute College, Washington, DC, June 26th, 27th, and 28th, 1889* (New York: Office of *The Deaf-Mutes' Journal*, 1890), 5.

275 *He suggested they*: Ibid.

276 *"I remain solitary"*: Alexander Graham Bell, letter to Mabel Hubbard Bell, May 5, 1890.

276 *Though Lucy had been*: Lucy Sanders, "How My Children Were Educated," *Silent Worker* 23, no. 10 (1911): 183.

276 *Lucy did the calculations*: Lucy Sanders, letter to Mrs. Sanders, undated, c. 1890.

276 *But there was no math*: Ibid.

277 *Standing before the gathering*: Alexander Graham Bell, *Marriage*, 3–4.

278 *A few years earlier*: Hiram Phelps Arms, *The Intermarriage of the Deaf: Its Mental, Moral and Social Tendencies* (Philadelphia: Burk & McFetridge, 1887), 34–35.

278 *Alec continued*: Alexander Graham Bell, *Marriage*, 4.

279 *Deaf people, however*: *Proceedings of the World Congress of the Deaf and the Report of the Fourth Convention of the National Association of the Deaf, held at the Memorial Art Palace*, Chicago, Ill. (1893), 114.

280 *Now Mabel said*: Bruce, *Bell*, 323; Mabel Hubbard Bell, letter to Alexander Graham Bell, October 4, 1893.

280 *Her life was now*: Mabel Hubbard Bell, letter to Eliza Symonds Bell, October 11, 1891; Mabel Hubbard Bell, letter to Eliza Symonds Bell, 1892; Marian (Daisy) Bell, memories of Mabel Hubbard Bell; Mabel Hubbard Bell, letter to Eliza Symonds Bell, October 23, 1892.

Chapter 18

283 *"Your deaf mute"*: Bruce, *Bell*, 308.

283 *Sarah Fuller held*: Keller, *Story of My Life*, 53; Michael Anagnos, "Helen Keller," *Perkins Annual Report* (1890): 120–26.

283 *The manual alphabet*: Ibid.; Herrmann, *Helen Keller*, 76.

284 *As different as Helen*: Keller, *Midstream*, 244–47. All of Helen's descriptions of meeting Bridgman are from this.

284 *It wasn't just Laura*: Helen Keller, "My Story," *The Youth's Companion*, 1895. Helen Keller Archive, American Federation of the Blind.

284 *"In the school where"*: Keller, *Story of My Life*, 42.

284 *In the year of*: Michael Anagnos, *Fifty-Sixth Annual Report of the Perkins Institution and Massachusetts School for the Blind* (Boston: Rand, Avery, & Co. 1888), 3. Disability History Museum.

285 *People were enthralled*: Joseph P. Lash, *Helen and Teacher: The Story of Helen Keller and Anne Sullivan Macy* (Boston: De Capo Press, 1997), 77–89.

285 *In 1890, after*: Keller, *Story of My Life*, 53.

285 *Alec had been following*: Sarah Fuller, letter to Alexander Graham Bell, May 3, 1890.

286 *The Volta Lab*: Marian (Daisy) Bell, "Alexander Graham Bell and the Volta Bureau," 2.

286 *The book was so*: *Helen Keller: Souvenir of the First Summer Meeting of the American Association to Promote the Teaching of Speech to the Deaf* (Washington, DC: Volta Bureau, 1892).

286 *At the Bells'*: Keller, *Midstream*, 111.

287 *Mostly, Helen spent*: Ibid.

287 *Still, Mabel*: Mabel Hubbard Bell, letter to Alexander Graham Bell, May 17, 1893.

287 *In 1895*: Amos Draper, "The Attitude of the Adult Deaf Towards Pure Oralism," *American Annals of the Deaf and Dumb* 40, no. 1 (January 1895): 53.

288 *What they did believe*: Ibid., 54.

288 *By 1895, manualists*: Edward Miner Gallaudet, "Methods of Instruction in American Schools," *American Annals of the Deaf and Dumb* 40, no. 1 (January 1895): 71–73.

289 *It hadn't always been*: *Proceedings of the Third National Convention of the Deaf* (New York: *National Deaf-Mutes' Journal*, 1889), 66, Collection of the National Association of the Deaf, Gallaudet University.

289 *"It asks all deaf"*: Draper, "The Attitude of the Adult Deaf," 51.

289 *"pure oralism makes"*: Ibid.

290 *Meanwhile, Alec was still*: Gardiner Greene Hubbard, letter to Alexander Graham Bell, July 2, 1895.

290 *Mabel, too, was worried*: Mabel Hubbard Bell, letter to Alexander Graham Bell, July 9, 1895.

291 *Around this time*: Ibid.

292 *But here at this*: Alexander Graham Bell, letter to Mabel Hubbard Bell, July 22, 1895. Emphasis in original.

292 *Officially, the CAID*: Edward Miner Gallaudet, "The Convention as Organized at Flint," *American Annals of the Deaf and Dumb* 41, no. 1 (1900): 13.

292 *By then, Helen was*: Nella Braddy Henney, *Anne Sullivan Macy: The Story Behind Helen Keller* (Garden City, NY: Doubleday, Doran, 1933), 167.

293 *There were about four hundred*: Helen Keller, *Story of My Life, with her letters (1887–1901) and a supplementary account of her education, including passages from the reports and letters of her teacher, Anne Mansfield Sullivan, by John Albert Macy* (Garden City, NY: Doubleday, Page & Co., 1913), 392.

293 *She didn't talk about*: Ibid.

293 *It wasn't simple*: Ibid.

295 *It's not as though*: Edward Allen Fay, "An Inquiry Concerning the Results of Marriages of the Deaf in America," *American Annals of the Deaf and Dumb*

42, no. 1 (1897): 29; Joseph C. Gordon, ed., *The Education of Deaf Children* (Washington, DC: Volta Bureau, 1892), 24.

296 *But Fay asks*: Edward Allen Fay, *Marriages of the Deaf in America: An Inquiry Concerning the Results of Marriages of the Deaf in America* (Washington, DC: Gibson Brothers, 1898), 27–28.

296 *"When opportunity occurs"*: Ibid., 29.

296 *"our friend"*: *Proceedings of the World's Congress of the Deaf and the Report of the Fourth Convention of the National Association of the Deaf, Held at the Memorial Art Palace* (Chicago, 1893).

Chapter 19

297 *"Jules Verne's books"*: Marian (Daisy) Bell, "Notes about Alexander Graham Bell."

297 *"The more I look at"*: Keller, *Midstream*, 132.

297 *In the five years*: Octavian E. Robinson and Jonathan Henner, "The Personal Is Political in *The Deaf Mute Howls*: Deaf Epistemology Seeks Disability Justice," *Disability & Society* 32, no. 9 (April 2017): 7, https://doi.org/10.1080/09687599.2017.1313723; Edward Allen Fay, "Tabular Statement of American Schools for the Deaf, 1899–1900," *American Annals of the Deaf and Dumb* 45, no. 1 (1900): 62–78.

298 *There was a time when*: Keller, *Story of My Life*, 31–33.

298 *And so Helen*: Ibid.

298 *Anne took her*: Ibid.

298 *Helen felt the meaning*: Ibid.

298 *Now, on Alec's*: Keller, *Midstream*, 132–34.

298 *They sat in silence*: Ibid.

299 *But Helen insisted*: Ibid.

299 *Alec did worry*: Ibid.

299 *All her life*: Nielsen, *The Radical Lives of Helen Keller* (New York: New York University Press, 2004), 128. "[Nella Braddy] Henney [Helen Keller's editor and author of *Anne Sullivan Macy*: *The Story Behind Helen Keller*] was determined that Keller be remembered as an inspirational and international figure of caring, the miraculous product of the highly skilled Anne Sullivan Macy. She devoted decades of her life to maintaining a rich intellectual and social circle for Keller, perhaps partially to fulfill a commitment to Anne Macy. By leaving this considerable collection of historical materials, Henney also guaranteed that her voice continues to be heard and her influence continues to hold sway—perhaps even to the exclusion of others."

299 *But connections to other*: Helen Keller, letter to Bertha Galeron de Calonne,

October 27, 1905, and March 31, 1906, American Foundation of the Blind, translated from the French by Martin Doppelt.

300 *Their friendship soon*: Helen Keller, letter to Bertha Galeron de Calonne, 1906, American Foundation of the Blind, translated from the French by Martin Doppelt.

300 *"I find myself"*: Ibid.

300 *"I wonder"*: Keller, *Midstream*, 255–59.

301 *Everywhere, science was*: Bruce, *Bell*, 305–06.

301 *By 1881*: Bruce, *Bell*, 292–93.

301 *"It is evident"*: *Proceedings of the Eighth Convention of the National Association of the Deaf*, 15–16.

302 *By then there was*: Ibid.; Lane, *When the Mind Hears*, 359; Greenwald and Van Cleve, " 'A Deaf Variety of the Human Race,' " 39.

302 *Meanwhile, marriage restriction*: Greenwald and Van Cleve, " 'A Deaf Variety of the Human Race,' " 39; Alexander Graham Bell, "A Few Thoughts Concerning Eugenics," *National Geographic*, February 1908, 122–23.

302 *"Among the unalienable rights"*: Bell, "A Few Thoughts Concerning Eugenics," 122–23.

302 *In 1905, sterilization*: The second was passed in 1909, in Washington, DC, followed by California and Connecticut in the same year. In 1911, Nevada, New Jersey, and Iowa followed suit. In 1912, New York joined them. For more information on Bell, eugenics laws, and how they affected the deaf, refer to research by Brian Greenwald, notably, " 'A Deaf Variety of the Human Race': Historical Memory, Alexander Graham Bell, and Eugenics," in *Journal of the Gilded Age and Progressive Era*, and "Revisiting the *Memoir*: Contesting Deaf Autonomy and the Real Tragedy of Alexander Graham Bell," in *In Our Own Hands: Essays in Deaf History, 1780–1970.*

302 *By 1909, Charles Davenport*: Greenwald and Van Cleve, " 'A Deaf Variety of the Human Race,' " 39–40.

303 *When Fay turned him*: Ibid.

303 *In 1912, ignoring*: Lane, *When the Mind Hears*, 359. It is notable that most states did not actually include the deaf in their sterilization laws. Further research on this is being conducted by Brian Greenwald.

303 *Robert Smithdas*: Personal interview, Richard McGann, February 27, 2020, as remembered to him by Robert Smithdas.

303 *"He lied"*: Ibid.

304 *In the landmark case*: Terry Gross, "The Supreme Court Ruling That Led to 70,000 Forced Sterilizations," *Fresh Air*, NPR, March 7, 2016.

304 *Soon after* Buck v. Bell: Mark Willis, "Not This Pig: Dignity, Imagination, and

Informed Consent," in *Genetics, Disability and Deafness*, ed. John Vickrey Van Cleve (Washington, DC: Gallaudet University Press, 2004), 174.

304 *In 1933, Hitler's*: John S. Schuchman, "Deafness and Eugenics in the Nazi Era," in *Genetics, Disability and Deafness*, ed. John Vickrey Van Cleve (Washington, DC: Gallaudet University Press, 2004), 72–77.

305 *And it worked*: Harlan Lane, *The Mask of Benevolence: Disabling the Deaf Community* (New York: Alfred A. Knopf, 1992), 113; Douglas C. Baynton, *Forbidden Signs: American Culture and the Campaign Against Sign Language* (Chicago: University of Chicago Press, 1996), 5.

306 *Veditz's body begins*: *The Preservation of Sign Language*, directed by George Veditz (1913), film; as translated by Carol Padden and Eric Malzkuhn, *Deaf World*, 83; observations of body language, pacing, emphasis, and nuance were made largely by Kenneth DeHaan on January 16, 2020, in a personal interview.

306 *He signs with pride*: Ibid.

306 *He knew that oralism*: Ibid. Emphasis in original translation.

306 *"These men are trying"*: Ibid.

306 *"As long as we have"*: Ibid.

307 *And even Mabel*: Mabel Hubbard Bell, letter to Gilbert Grosvenor, October 11, 1921.

308 *"I have helped other things"*: Mabel Hubbard Bell, letter to Gilbert Grosvenor, October 11, 1921.

309 *She had been thinking*: Mabel Hubbard Bell, letter to Alexander Graham Bell, January 12, 1898; Toward, *Mabel Bell*, 168–69.

309 *When Gertrude died*: Yale, *Years of Building*, 237–40.

309 *Mabel had shown*: Toward, *Mabel Bell*, 240; Mabel Hubbard Bell, letter to Elsie Bell, October 12, 1921.

310 *Alec never did*: Marian (Daisy) Bell, memories of Alexander Graham Bell, written on the train, July 4, 1926.

310 *He didn't like doctors*: Marian (Daisy) Bell, "Notes about Alexander Graham Bell." "There has never been any doubt in my mind but that Father would have lived longer had he consented to see a doctor and obey him," wrote Daisy.

310 *"a creature of the water"*: Marian (Daisy) Bell, memories of Alexander Graham Bell, written on the train, July 4, 1926.

310 *In his last*: Mabel Hubbard Bell, *Home Notes*, 134, 136.

311 *July 22 was*: Ibid., 115.

311 *The next day*: Ibid., 118.

311 *By July 30*: Ibid., 120.

311 *Mabel and Alec*: Ibid., 119, 136.

311 *When he woke*: Ibid., 136.
312 *The several times*: Ibid., 136.
312 *Around six*: Ibid., 101, 125.
312 *Even then, for a moment*: Ibid., 119.

Afterword

315 *"Oralists in their efforts"*: Ballin, *The Deaf Mute Howls*, 26.
317 *Wyatte Hall, a deaf*: Wyatte Hall, personal interview, September 2017.
318 *In health-care settings*: Wyatte C. Hall, Leonard L. Levin, and Melissa L. Anderson, "Language Deprivation Syndrome: A Possible Neurodevelopmental Disorder with Sociocultural Origins," *Social Psychiatry and Psychiatric Epidemiology* 52, no. 6 (June 2017): 761–76; Neil Glickman, "Do You Hear Voices? Problems in Assessment of Mental Status in Deaf Persons With Severe Language Deprivation," *Journal of Deaf Studies and Deaf Education* 12, no. 2 (Spring 2007):127–47.
318 *These barriers extend*: McCay Vernon, "The Horror of Being Deaf and in Prison," *American Annals of the Deaf and Dumb* 155, no. 3 (2010): 311–21; Sara Nović, "Deaf Prisoners Are Trapped in Frightening Isolation," *CNN Opinion*, June 21, 2018.
319 *In fact, many of the*: Matthew L. Hall, Inge-Marie Eigsti, Heather Bortfeld, and Diane Lillo-Martin, "Auditory Deprivation Does Not Impair Executive Function, but Language Deprivation Might: Evidence from a Parent-Report Measure in Deaf Native Signing Children," *Journal of Deaf Studies and Deaf Education* 22, no. 1 (January 2017): 9–21; Nils Skotara, Uta Salden, Monique Kügow, Barbara Hänel-Faulhaber, and Brigitte Röder; "The Influence of Language Deprivation in Early Childhood on L2 Processing: An ERP Comparison of Deaf Native Signer and Deaf Signers With a Delayed Language Acquisition," *BMC Neuroscience* 13, no. 1 (May 2012): 13–44.
319 *In the hearing world*: Sanjay Gulati, "Language Deprivation Syndrome."
320 *In 1907, when George*: Lane, *When the Mind Hears*, 340.
320 *Like oralism*: Hall, Levin, and Anderson, "Language Deprivation Syndrome."
320 *Oralism and LSL*: It is also sometimes shortened to AGB.
321 *If Bell was hated*: Hall, Levin, and Anderson, "Language Deprivation Syndrome."
322 *In a 2017 paper*: Robinson and Henner, "The Personal Is Political," 4, https://doi.org/10.1080/09687599.2017.1313723.
325 *Students spilled out*: Douglas Baynton, Jack R. Gannon, and Jean Lindquist Bergey, *Through Deaf Eyes: A Photographic History of an American Community* (Washington, DC: Gallaudet University Press, 2007), 132.

326 *Oliver Sacks, who was*: Oliver Sacks, *Seeing Voices: A Journey into the World of the Deaf* (New York: Vintage Books, 2000), 130.

326 *rarely did anyone*: The significant exception is Ballin, whom Bell made to feel stupid when he was trying to defend oralism. (Ballin, *The Deaf Mute Howls*, 42).

Selected Bibliography

Arms, Hiram Phelps. *The Intermarriage of the Deaf: Its Mental, Moral and Social Tendencies.* Philadelphia: Burk & McFetridge, 1887.

Ballin, Albert. *The Deaf Mute Howls.* Washington, DC: Gallaudet University Press, 1998.

Baynton, Douglas C. *Forbidden Signs: American Culture and the Campaign Against Sign Language.* Chicago: University of Chicago Press, 1996.

Baynton, Douglas C., Jack R. Gannon, and Jean Lindquist Bergey. *Through Deaf Eyes: A Photographic History of an American Community.* Washington, DC: Gallaudet University Press, 2007.

Bell, Alexander Graham. *Marriage: An Address to the Deaf.* 3rd ed. Washington, DC: Sanders Printing Office, 1898.

———. *Memoir upon the Formation of a Deaf Variety of the Human Race.* N.p.: Alexander Graham Bell Association for the Deaf, 1969.

———. *The Multiple Telegraph: Invented by A. Graham Bell.* Boston: Rand, Avery, & Co.1876.

———. *The Question of Sign-Language and the Utility of Signs in the Instruction of the Deaf, Two Papers.* Washington, DC: Sanders Printing Office, 1898.

———. *Upon a Method of Teaching Language to a Very Young Congenitally Deaf Child.* Washington, DC: Gibson Brothers, 1883.

Bell, Alexander Melville. *Visible Speech: A New Fact Demonstrated.* London: Hamilton, Adams,1865.

———. *Visible Speech: The Science of Universal Alphabetics; or Self-Interpreting Physiological Letters, for the Writing of All Languages in One Alphabet.* London: Simpkin, Marshall & Co, 1867.

Bell, Charles David, and Alexander Melville Bell. *Bell's Standard Elocutionist.* London: William Mullan & Son, 1878.

The Bell Telephone: The Deposition of Alexander Graham Bell in the Suit Brought by the United States to Annul the Bell Patents. Boston: American Bell Telephone Company, 1908. Massachusetts Historical Society.

Biesold, Horst. *Crying Hands: Eugenics and Deaf People in Nazi Germany*. Translated by William Sayers. Washington, DC: Gallaudet University Press, 1999.

Boatner, Maxine Tull. *Voice of the Deaf: A Biography of Edward Miner Gallaudet*. Washington, DC: Public Affairs Press, 1959.

Bruce, Robert V. *Alexander Graham Bell: Teacher of the Deaf*. Northampton, MA: Clarke School for the Deaf, 1974. Massachusetts Historical Society.

———. *Bell: Alexander Graham Bell and the Conquest of Solitude*. Boston: Little, Brown, 1973.

Deaf-Mute Education in Massachusetts: Report of the Joint Special Committee of the Legislature of 1867 on the Education of Deaf-Mutes with an Appendix Containing Evidence, Arguments, Letters, etc., Submitted by the Committee. Boston: Wright & Potter, 1867.

DeLand, Fred. "Biography of Sarah Fuller." Unpublished manuscript. Alexander Graham Bell Family Papers, Library of Congress.

———. *Dumb No Longer: The Romance of the Telephone*. Washington, DC: Volta Bureau, 1908.

———. "A Short Biography of Alexander Graham Bell." Unpublished manuscript and notes. Alexander Graham Bell Family Papers, Library of Congress.

Fay, Edward Allen. *Marriages of the Deaf in America: An Inquiry Concerning the Results of Marriages of the Deaf in America*. Washington, DC: Gibson Brothers, 1898.

Gannon, Jack R. *Deaf Heritage: A Narrative History of Deaf America*. Silver Spring, MD: National Association of the Deaf, 1981.

Gray, Charlotte. *Reluctant Genius: Alexander Graham Bell and the Passion for Invention*. New York: Arcade, 2011.

Greenwald, Brian H., and John Vickery Van Cleve, eds. *A Fair Chance in the Race of Life: The Role of Gallaudet University in Deaf History*. Washington, DC: Gallaudet University Press, 2008.

Greenwald, Brian H., and Joseph J. Murray. *In Our Own Hands: Essays in Deaf History, 1780–1970*. Washington, DC: Gallaudet University Press, 2016.

Grosvenor, Edwin, and Morgan Wesson. *Alexander Graham Bell: The Life and Times of the Man Who Invented the Telephone*. New York: Harry Abrams, 1997.

Henney, Nella Braddy. *Anne Sullivan Macy: The Story Behind Helen Keller*. Garden City, NY: Doubleday, Doran, 1933.

Hitz, John. *Dr. A. Graham Bell's Private Experimental School*. Washington, DC: Sanders Printing Office, 1898.

Hubbard, Gardiner G. *Addresses Delivered at the Twenty-Fifth Anniversary of the Opening of the Clarke Institution, Northampton, Mass., October 12, 1892*. Northampton, MA: Gazette Printing, 1893.

———. *The Story of the Rise of the Oral Method in America, as Told in the Writings of the Late Hon. Gardiner G. Hubbard.* Washington, DC: Press of W. F. Roberts, 1898.

Jaeger, Paul T., and Cynthia Ann Bowman. *Disability Matters: Legal and Pedagogical Issues of Disability in Education.* Westport, CT: Bergin & Garvey, 2002.

Keller, Helen. *Midstream: My Later Life.* Garden City, NY: Sun Dial Press, 1937.

———. *The Story of My Life.* New York: W. W. Norton, 2004.

Krentz, Christopher, ed. *A Mighty Change: An Anthology of Deaf American Writing, 1816–1864.* Washington, DC: Gallaudet University Press, 2000.

Lane, Harlan. *The Mask of Benevolence: Disabling the Deaf Community.* New York: Alfred Knopf, 1992.

———. *When the Mind Hears: A History of the Deaf.* New York: Random House, 1984.

Lane, Harlan, Robert Hoffmeister, and Ben Bahan. *A Journey into the Deaf-World.* San Diego: DawnSignPress, 1996.

Lash, Joseph P. *Helen and Teacher: The Story of Helen Keller and Anne Sullivan Macy.* Boston: De Capo Press, 1997.

Longmore, Paul K., and Lauri Umansky, eds. *The New Disability History: American Perspectives.* New York: New York University Press, 2001.

Mackenzie, Catherine. *Alexander Graham Bell: The Man Who Contracted Space.* Boston, New York: Houghton Mifflin, 1928.

Mann, Horace. *Seventh Annual Report of the Board of Education* [of Massachusetts]. Boston: Dutton and Wentworth, 1844.

Millard, Candice. *Destiny of the Republic: A Tale of Madness, Medicine and the Murder of a President.* New York: Doubleday, 2011.

Nielsen, Kim E. *Beyond the Miracle Worker: The Remarkable Life of Anne Sullivan Macy and Her Extraordinary Friendship with Helen Keller.* Boston: Beacon Press, 2009.

———. *The Radical Lives of Helen Keller.* New York: New York University Press, 2004.

Reé, Jonathan. *I See a Voice: Deafness, Language and the Senses—A Philosophical History.* New York: Metropolitan Books, 1999.

Ryan, Donna F., and John S. Schuchman, eds. *Deaf People in Hitler's Europe.* Washington, DC: Gallaudet University Press, 2002.

Sacks, Oliver. *Seeing Voices: A Journey into the World of the Deaf.* New York: Vintage Books, 2000.

Shulman, Seth. *The Telephone Gambit: Chasing Alexander Graham Bell's Secret.* New York: W. W. Norton, 2008.

Speech for the Deaf: Essays Written for the Milan International Congress: Proceedings and Resolutions. London: W. H. Allen, 1880.

Standage, Tom. *The Victorian Internet: The Remarkable Story of the Telegraph and the Nineteenth Century's On-Line Pioneers.* New York: Walker, 1998.

Toward, Lilias M. *Mabel Bell: Alexander's Silent Partner.* Wreck Cove, Nova Scotia: Breton Books, 1996.

Van Cleve, John Vickrey, and Barry A. Crouch. *A Place of Their Own: Creating the Deaf Community in America.* Washington, DC: Gallaudet University Press, 1989.

Waite, Helen Elmira. *Make a Joyful Sound: The Romance of Mabel Hubbard and Alexander Graham Bell.* Philadelphia: Macrae Smith, 1961.

Watson, Thomas. *The Birth and Babyhood of the Telephone.* N.p.: American Telephone and Telegraph Company, 1913. Massachusetts Historical Society.

———. *Exploring Life: The Autobiography of Thomas A. Watson.* New York: D. Appleton, 1926.

Winefield, Richard. *Never the Twain Shall Meet: Bell, Gallaudet, and the Communications Debate.* Washington, DC: Gallaudet University Press, 1987.

Yale, Caroline A. *Years of Building: Memories of a Pioneer in a Special Field of Education.* New York: Dial Press, 1931.

Index